Of Plants and People

Of Plants and People

BY CHARLES B. HEISER, JR.

UNIVERSITY OF OKLAHOMA PRESS : NORMAN

BY CHARLES B. HEISER, JR.

Nightshades: The Paradoxical Plants (San Francisco, 1969)
Seed to Civilization: The Story of Food (San Francisco, 1973; 1981)
The Sunflower (Norman, 1976)
The Gourd Book (Norman, 1979)

Library of Congress Cataloging in Publication Data

Heiser, Charles Bixler, 1920–
  Of plants and people.

  Bibliography: p.
  Includes index.
  1. Plants, Cultivated—America. 2. Botany, Economic—Amer-
ica. 3. Ethnobotany—America. 4. Plants, Cultivated—America
—Folklore. I. Title.
SB87.A45H45   1985   581.6'09181'2   84-28126
ISBN 0-8061-1931-4

The paper in this book meets the guidelines for permanence
and durability of the Committee on Production Guidelines for
Book Longevity of the Council on Library Resources, Inc.

To Jorge Soria, Sergio Soria, and Jaime Díaz,
who introduced me to the beautiful country of Ecuador

Contents

vii

Illustrations

Preface

In the last few years at least three people have asked me when I was going to write another book. It is that overwhelming popular demand that has led to this volume. I must also admit that some of the kind remarks of the reviewers of my other books, particularly those by the distinguished botanist Herbert Baker (in *Pacific Horticulture*, Spring, 1982), have served as a stimulus to continue writing.

Unlike my other works, this one is not confined to one subject or to a single group of plants. Rather, for the most part it is the stories of miscellaneous plants. The plants are all American, most of them being from the Andes, and the unifying thread, if there is one, is that all of the plants have interesting interactions with people (at least I think they are interesting, but that will be up to the reader to decide). I have been close to most of the plants that I write about through either my own research or that of my students. The last chapter concerns religion and the origin of agriculture, and in its writing it was necessary that I wear another hat. Speculation runs rampant on that subject, and how much of that chapter has been done tongue-in-cheek I shall let the readers decide. Some parts of this book have been published in other forms in various journals, but large parts of it are previously unpublished, and some new interpretations are offered.

The book is intended for anyone who is interested in plants and people. That audience, I hope, will include not only botanists, horticulturists, and anthropologists but nonprofessionals as well. I have tried to write with the student and amateur in mind, and I have used as little botanical jargon as possible. There is very little material

of a difficult nature and practically no chemistry or mathe-
matics, subjects so important to modern biology. The book
is pretty much old-fashioned botany, perhaps because I am
pretty much an old-fashioned botanist. At times, I admit,
I get carried away in the discussion of the origin, development, and improvement of the domesticated plants, but
these are subjects of special interest to me, and I only
hope I don't run them into the ground. After all, in order
to continue to feed itself, humankind must depend upon
the improvement of our domesticated plants, and I feel
that an understanding of their origins may sometimes
contribute to that end. Although the plants considered
here are rather minor ones, some of them could well make
more significant contributions to our foods in the future.
Two of them, *quinua* and amaranth, have already been
recognized as having a great potential, but another one,
sumpweed, presently is not even a cultivated plant.

On rereading the manuscript I am embarrassed to see
the word *I* so often (but not embarrassed enough to change
it). It is repeated so many times not to try to indicate any
great importance of my work, but to attempt to convey
some of my enthusiasm for working with plants. After all,
this is a personal account.

References are given at the end of the book. These are
not intended to be complete, but anyone interested in
pursuing any topic in depth will find additional references
in most of the works cited.

Except where otherwise indicated, all of the photographs
are my own. Many of these have been converted to black
and white from Kodachromes, and some did not come out
as well as I would have liked. On my early trips to the
Andes I took only Kodachromes, and it has not been possible to duplicate some of the photographs in black and
white on my later trips.

The people who have helped in one way or the other in
gathering the material for this book are numerous, and it
will not be possible to mention all of them. Particular
thanks go to my former students and to many people in

Latin America, many of whom are mentioned in the book. Others include Guido and Modesto Soria, Carlos Ochoa, Francisco Vivar, and Miguel Holle. Special thanks go to Lewis Johnson and Donald Burton for reading the manuscript; to Virginia Flack, who typed it (she is probably the only person in the world who can read my longhand); and, of course, to my wife, Dorothy, who accompanied me on many of the trips.

Of Plants and People

CHAPTER 1

Of Pepos and People

Halloween or All Hallows can be traced to the autumnal festival of the Celts. Whatever the original purpose was—perhaps a pastoral and harvest celebration related to fertility—the celebration early acquired sinister implications. The dead were supposed to return home on that day, and witches, ghosts, and hobgoblins were thought to roam the fields. Fire rites, divination—bobbing for apples was once used to determine one's future spouse—and masquerades became associated with Halloween as time passed. How early jack-o'-lanterns became involved is not known, but we can be fairly certain that pumpkins were not used at first, for they are American plants, unknown to Europe until after the time of Columbus. Hollowed-out turnips may have served as the first jack-o'-lanterns. Who Jack was is also far from certain, but an Irish legend has it that there was a man named Jack, who, forbidden to enter Heaven because of his stinginess and barred from Hell because of his practical jokes, was condemned to walk the earth with his lantern until Judgment Day.

Jack-o'-lanterns are, of course, along with trick-or-treating, a traditional part of Halloween observances today in the United States. Some youngsters, for reasons hard to fathom, also have made the stealing of pumpkins from doorsteps and the squashing of them in the street a part of the celebration, and one can only hope that they will slip in the squashed pumpkin. The average family probably buys its only pumpkin at Halloween for making the jack-o'-lantern. Pumpkins, of course, have not disappeared as food, for the pumpkin is still traditional with many families for Thanksgiving pies, and they are found in cafe-

A typical sight in late November in much of the United States—Halloween pumpkins for sale.

terias and bakeries most of the year. However, if the average American family makes a pumpkin pie, it probably buys the pumpkin in a can. Today pumpkins are much less used for human food than they were at one time, for pumpkin, along with squash, was probably the most important food, after corn and beans, of the Indians of eastern North America. If we are to believe a rhyme from New England, our ancestors ate a lot more pumpkin than we do.

For pottage and puddings and custards and pies,
Our pumpkins and parsnips are common supplies;
We have pumpkin at morning and pumpkin at noon;
If it were not for pumpkins we should be undone.

The pumpkin "pie" of colonial New England was very

different from that of today. An old recipe tells us that it was made by cutting a hole in the pumpkin to permit the removal of the seeds, after which it was stuffed with apples, spices, sugar, and milk and then baked.

As everyone knows, it is pumpkins, and not squashes, that are associated with Halloween. That is, everyone knows but Marcie of the "Peanuts" comic strip. All regular followers of "Peanuts" know that every Halloween, Linus waits up for the Great Pumpkin to appear. In the episode on the day after Halloween in 1973, Marcie asked Linus if the "Great Squash" ever showed up, and in the final frame he yelled at her, "That's PUMPKIN." All of this is by way of asking what is the difference between a squash and a pumpkin?—a very good question, as people are wont to say to gain time to think of an answer. Botanically speaking, there is really no basic difference. All squashes and pumpkins belong to the genus *Cucurbita*, and more than one species contain fruits called squashes and pumpkins as well as gourds. Perhaps the best known to most of my readers are those that are found in the species *Cucurbita pepo*,[1] which includes the Halloween pumpkin and the pie pumpkin, several summer squashes (such as zucchini, patty-pan, and yellow crookneck), and the acorn winter squash, as well as ornamental gourds. The pumpkins, according to my definition, are generally round and orange, whereas squashes are not. The chief difference, however, seems to come from the way they are used in the kitchen: usually we eat squashes as a vegetable and pumpkins as a dessert, but I should point out that some squashes can also be used to make a good pie. Marcie, it turns out, was really not too far off the mark.

In the fall of the year, newspapers frequently publish a

---

[1] This scientific name has also made the comic strips. In "Our Boarding House" for November 28, 1973, Major Hoople is taking a stroll and sees a stand with pumpkins for sale, whereupon he says, "Ha. What wonderful examples of *Cucurbita pepo*, the traditional symbol of Halloween." I tell my students that if Major Hoople can remember the scientific name, I expect them to be able to do so.

photograph of a large pumpkin grown by Mr. or Mrs. So-
and-so, and contests are held for the largest pumpkins or
squashes in various parts of the country. I am not sure of
the current record, but some pumpkins weigh over five
hundred pounds, truly remarkable fruits—without doubt
the largest fruits known in the plant kingdom. Are these
gigantic fruits pumpkins or squashes? Insofar as I can
judge from a photograph, I think the largest belong to
a variety of *Cucurbita maxima,* which is generally considered a squash. On occasion I have had people ask me how
to grow one of these gigantic fruits. Although I am not a
specialist on the subject, I tell them that the first thing
to do is to start with seeds from a giant pumpkin or squash.
Some seed companies specialize in these. Next, they should
make sure they start the seeds early, usually in the house,
and then grow them under ideal conditions—a very fertile
soil, lots of sunlight, and plenty of water—and probably
they should allow only one fruit to develop on the vine.
As I recall from the stories of Laura Ingalls Wilder, which
I read to my children a long time ago, she once fed milk
to the vine in order to get an extra large pumpkin. I have
not looked into how—or if—that works, for it is not one of
my ambitions to grow the world's largest pumpkin.

In recent years squashes have increased in popularity.
In fact, there is a whole cookbook devoted to one of them,
the zucchini. Some people perhaps are unaware that the
flowers of squashes and pumpkins are a popular food in
some places. In many markets of Mexico one may see large
stacks of the flowers. No fruits are sacrificed as a result,
for a squash plant produces a large number of male flowers
(or, for the purists, I should say staminate flowers) in pro-
portion to the number of female (pistillate) flowers, and
generally only the former are used. I have eaten them in
quesadillas, which I found rather tasty, although I am not
sure that the flowers added much to the flavor. The seeds
of squashes and pumpkins are also eaten and may be obtained in many places, and in Mexico City, street-corner
vendors selling *pepitas* are a common sight. It has been

A pumpkin vine takes over a home garden.

postulated that the seeds may have been the first part of the plant eaten, for the gourds that are the likely ancestors of the squashes and pumpkins have little or no flesh, and what may be present is extremely bitter. Mutants giving more flesh and a loss of the bitterness may have occurred after the plant was cultivated for its seeds, and in time the flesh became more important than the seeds. The latter, however, are a better food in that they contain a large amount of oil and fair amounts of protein.

Pumpkins, squashes, and other cucurbits have not only entered our stomachs, but also have entered our language in ways other than to refer to the plants or their fruits. They have been used to denote stupidity (pumpkin- or gourd-head for a dull or stupid person) and because of their shape or size often have sexual connections (supply your own example). Such matters have been thoroughly explored in a book that appeared in 1980 with the impressive title *Nature and Language: A Semiotic Study of Cucurbits in Literature.* Semiotics is defined as a general theory of signs and symbolism, and the reader will learn that various cucurbits have long functioned as signs or symbols. Although that book is a serious work, I found it both entertaining and humorous. Those interested in cucurbits will perhaps find the first part on "pregnant gourds" and "delirious pumpkins" the most fascinating, but for the student of semiotics the chapters on "Is a library work written by the author or by the readers?" and "Does a literary work write itself?"[2] may be more appealing.

The authors of *Nature and Language,* Ralf Norrman and Jon Haarberg, did not find it necessary to consider chronology and history for their purposes, but they do point out that cucurbitic references increase rapidly in the sixteenth century. This is not surprising, for history tells us that that was the time when American cucurbits, particularly the squashes and pumpkins, became known in Europe. The word *squash* comes from an American Indian word, so it could not have been used before that time. *Pumpkin,* however, is apparently derived from the Greek word πέπων, "pepon" (or *pepo* in Latin, now used to designate the specialized berry or type of fruit found in most cucurbits). This word became attached to the American plant, becoming *pumpion* or *pompion* in early English writings before arriving at its present form. The scientific names of these plants, as well as the common names, are also sometimes the source of confusion. Although today the genus name *Cucurbita* is used exclusively for plants that were native to the Americas (and it has only been fairly recently established that they all are American), the word *cucurbita* is the Latin for gourd. The reason for some confusion is that when Linnaeus established the genus *Cucurbita* over two hundred years ago, he included the bottle gourd (now

Lagenaria siceraria) and the watermelon (now *Citrullus lanatus*), both of Old World origin, as well as some of the pumpkins and squashes.

The word *cucurbitare,* from a Latin work of the twelfth century, can be added to the list of cucurbitic words compiled by Norrman and Haarberg, and according to F. Deltgen and H. G. Kauer, it means "to commit adultery." From their explanation we learn that "in particular, it refers to a vassal who has seduced the wife of his feudal lord and who in this way makes her abdomen swell like a pumpkin, i.e., he makes her pregnant." To quibble with this explanation, I might point out that it would be better to use *gourd* instead of *pumpkin,* in view of the fact that the pumpkin, as now recognized, was unknown to the Romans.

As a boy I used to visit my grandfather's farm in southern Indiana in the summer. It was here that I first heard about "contamination" in cucurbits, although I don't recall that this word was used. "Contamination" supposedly occurs when cantaloupes grow near cucumbers or squash grows near gourds, and as a result the cantaloupes or squash have a bitter taste and are unfit to eat. I am not sure that my grandfather believed this, but as I recall he took no chances and grew his cucumbers some distance from cantaloupes. (Actually, he grew another melon, but they are called cantaloupes in Indiana.)

The earliest account of "contamination" of which I am aware comes from Cotton Mather, who in 1716 wrote of a friend of his near Boston whose garden was robbed of squashes on occasion:

> To inflict a pretty little punishment on the Theeves, he planted some *Guords* among the Squashes, (which are in aspect very like 'em) at certain places which he distinguished with a private mark, that he might not be himself imposed upon. By this method, the Thieves were deceived, & discovered, & ridiculed. But yet the honest man saved himself no squashes by ye Trick; for they were so infected and Embitter by the Guords, that there was no eating of them.

Also, in a letter from "J. B." (possibly John Bartram) in *Gentlemen's Magazine* in 1755 one reads: "If we plant cucumbers, squashes, or melon, near the bitter gourd, the fruits of the first will be as bitter as gall . . . this shows how liable plants are to be bastardized by bad neighbors."

Another account is given by Asa Gray in 1857 in his review of Naudin's monograph of *Cucurbita*. After pointing out that Naudin found that the different species of *Cucurbita* will not hybridize, he writes:

What are we to think, then, of the universal belief that squashes are spoiled by pumpkins grown in their vicinity, or pumpkins by squashes; and even melons (which are of a different genus) by squashes? The fact of some such influence seems to be well authenticated. Dr. Darlington, one of the most trustworthy of observers, speaks of it from his own knowledge, thus: "When growing in the vicinity of squashes the fruit [of the pumpkin] is liable to be converted into a kind of hybrid of little or no value. I have had a crop of pumpkins totally spoiled by that cause, the fruit becoming very hard and warty, unfit for the table and unsafe to give to cattle."

Now that this is not the effect of hybridation [*sic*] is clear from the fact that the result appears in the fruit of the season, not in that of the next year, namely, in a generation originated by the crossing. A clue is perhaps furnished by Naudin's observations, that the ovary is apt to set and even develop into a fruit in consequence of the application of the pollen of another species, although, as the result proves, none of the ovules are fertilized.

And he hazards the conjecture that the pollen may exert a specific influence first upon the ovary, inciting its farther development, and then upon the ovules. To test this conjecture he was to examine the action, if any there be, of the pollen of Cucurbita upon the ovary of melons.

Insofar as I can determine, Naudin never published any additional observations, or if he did, I haven't been able to find them. If Asa Gray, a noted botanical authority, had not taken the subject so seriously, I would be inclined to dismiss "contamination" as folklore. Two possible explanations occurred to me. First, in squashes, at least, it may be a result of hybridization, for both the ornamental yellow-flowered gourd and some pumpkins will hybridize with many of the most commonly grown squashes. If this were true, as Gray noted, the effect wouldn't be noticed until the year following the hybridization. However, if people found their squashes to be bitter as a result of hybridization, they might credit it to "contamination" occurring in the same year, not realizing that their seed were of hybrid origin. The other possibility is that if a cantaloupe were pollinated by a cucumber or a squash by a gourd, the pollen would affect the taste of the fruit. The pollen is known to have an effect on the fruit in some plants (the botanical term is *xenia*), but I could not find that any such effect had been recorded for cucurbits. So I decided to write to my good friend Thomas W. Whitaker, who is one of the world's authorities on cucurbits. He replied that the only scientific explanation is that cucumbers may emit volatile substances that are quickly absorbed by melons' fruits and which might give those fruits a cucumberish odor or flavor. He didn't comment on squashes, gourds, and pumpkins, but I imagine that the same might be said for them. He closed, however, by saying that this idea has never been investigated and "personally, I don't think it has substance."

After I had finished writing the preceding paragraphs, I wondered why no one had ever run an experiment to test for contamination. Perhaps that was because it wasn't terribly important, but I could think of many experiments on even less important subjects. Surely someone must have experimented for contamination, but perhaps the results were negative and were not thought worthy of publication, or perhaps the results were buried in some journal that had escaped my attention. Rather than conduct a more exhaustive search, I decided to conduct a little experiment myself as time permitted. In both the garden and the greenhouse I grew a summer squash next to an ornamental gourd and a cantaloupe next to a cucumber. At the same time, I grew a summer squash and a cantaloupe well isolated from other cucurbits. As the fruits matured, I had friends conduct taste tests and they could detect no differences between the fruits from the plants grown in isolation and those from the plants grown next to their "bitter" relatives. I went so far as to rub pollen of a cucumber on the stigma of a cantaloupe, and I did the same with the gourd and the squash. Both flowers produced fruits, which, of course, could have been from self-pollination, for they weren't bagged, and these fruits showed no bitterness. Later I grew plants in a greenhouse free of pollinating insects, and I pollinated a squash flower with gourd pollen. Again, the taste of the fruit was no different from those of the squashes secured by self-pollination. That is as far as the experiment went, and on the basis of it I am inclined to believe that contamination is a myth, but I am sure that some people will continue to plant their cantaloupes and squashes some distance from their cucumbers and gourds.

Five different species of *Cucurbita* were domesticated. In addition to *Cucurbita pepo*, there are three other annual species—*C. moschata* (our common types are butternut squash and Kentucky field pumpkin), *C. maxima* (Hubbard, buttercup, and turban squashes), and *C. mixta* (the cushaws)—and one perennial species, *C. ficifolia.* So far as I am aware the last is not grown in the United States for food but is sometimes grown as a novelty under the names of Malabar gourd and fig-leaf gourd. It is the only species adapted to high altitudes, and it is an important food in Mexico and highland South America.

In view of the great variability and the parallel development of the fruits in the various species, one may well ask how one distinguishes one species from another. It is not always easy, but if one has the whole vine, a few helpful characters in the leaves and stems can be found. It is possible to identify the species if one has only the fruit as long as the stalk (peduncle) is still attached, for it offers some of the best characters. The seeds also show slight, but supposedly consistent, differences among the species. Finally, the fruit flesh of some species is coarse-grained and of others more finely grained.

Most of the species appear to have been domesticated in

Mexico or Central America, but at least one of them, *Cucurbita maxima,* had its origin in South America. Many of the species figure prominently in the earliest agriculture in the Americas, and *C. pepo* is one of our oldest domesticated plants, perhaps going back to 5000 B.C. in Mexico.

Although these species have received considerable study by botanists, wild ancestral types are not definitely known for any of them. This is somewhat surprising, for the wild progenitors have now been identified for most of our major crop plants,and many of the minor ones as well. Two wild cucurbits, however, seem possible candidates. A naturally occurring gourd of Argentina and Bolivia, *Cucurbita andreana,* may be the wild ancestor of the domesticated *C. maxima,* but some have thought that it may be nothing more than an escaped form of it instead of a truly wild plant. A somewhat similar situation exists between *C. pepo* and the plant generally called *C. texana,* and the relation between the two calls for careful examination, particularly in view of a possible connection of *C. pepo* with the origin of agriculture in the eastern half of the United States.

Agriculture generally is thought to have developed much later in eastern North America than it did in Mexico, and corn, beans, squashes and pumpkins were found to be the principal crops of the Indians when North America became known to the Europeans. Archaeological investigations, however, reveal that two native plants, the sunflower and the sumpweed, were domesticated in the eastern United States, and certain other ones—lamb's-quarters, May-grass and a knotweed—may have been cultivated at one time. These plants were being used by the Indians before they had corn, which certainly came to them from Mexico. It has thus been suggested that agriculture had an independent origin in the eastern United States based on the native plants. The people may have acquired the idea of agriculture from Mexico, but they could have started agriculture with indigenous plants rather than ones borrowed from Mexico.

Recent archaeological evidence has thrown new light on the subject. First was the discovery of our old friend *Cucurbita pepo* in an archaeological site at Phillips Spring, Missouri, which was dated at 2000 B.C. Then, in 1983, material of this species from a site in west central Illinois was assigned a date of 5000 B.C. The material from Missouri consists of both rind fragments and seeds. The latter are rather small for a squash, about 10.5 mm (less than ½ inch) long, but they are larger than gourd seeds, so it could well be that they represent a primitive domesticated plant. If so, it would make squash the oldest domesticated plant known from the United States. Only rind fragments were found in Illinois, and it seems most likely that they come from a wild plant. Inasmuch as *C. pepo* is an old domesticated plant in Mexico, one explanation of the material from Missouri could be that it, or seeds for it, came from Mexico, but why would a wild gourd be found in Illinois at such an early date?

The presence of *Cucurbita texana*, the Texas gourd, in the United States may provide an answer to that question. It may also offer another explanation for a domesticated form of *C. pepo* in Missouri at 2000 B.C.: namely, that the squash was independently domesticated in both Mexico and central North America. A search of literature revealed that the idea was not original with me, for an independent domestication of *C. pepo* in eastern North America was postulated more than thirty years ago by Thomas Whitaker and George Carter. However, I did not find that either of them maintained that view, and in a paper in 1975, Whitaker gives a figure showing *C. pepo* coming to the eastern United States directly from Mexico, and he has also on occasion mentioned that there is no critical evidence to determine whether *C. texana* represents a truly wild species or a naturalized one.

If it is naturalized, that means, of course, that it is derived from *Cucurbita pepo* and not the reverse. A domesticated organism may sometimes escape from human control, revert to the wild state, and become well established.

The Texas gourd growing in the greenhouse. The fruit is about 3½ inches long.

A familiar example is the horse in the western United States. Sometimes it is most difficult to determine, however, if a particular organism is an escape that has become naturalized or is truly wild. With the horse, the answer is clear, for there were no horses living in the Americas before their introduction by the Spanish.

The idea that *Cucurbita texana* might be an escape originated with Asa Gray. The plant was first described as belonging to another genus in 1848, and two years later it was transferred to *Cucurbita* by Gray, who remarked, "the fruit is just that of *Cucurbita ovifera* (now *Cucurbita pepo* var. *ovifera,* the ornamental gourd), of which our plant might possibly be only a naturalized variety." This statement seems to be the sole basis for considering it a naturalized plant, and generally overlooked is Gray's later statement (1857) that the plant had the appearance of being native: "At least, this is the opinion of Mr. Lindheimer and Mr. Charles Wright [both well-known botanical collectors], two good judges. The latter informs us that, from the stations and localities in which alone it is met with, he could not so feel it to be other than an indigenous plant."

This latter opinion has been expressed by all botanists who have given careful study to this plant. In the late 1920's, Liberty Hyde Bailey, one of the foremost authorities on cultivated plants and a specialist in cucurbits, visited the area where *Cucurbita texana* grows and along with his host, the Texas botanist B. C. Tharp, concluded that *C. texana* was probably an indigenous plant. In 1943, Bailey stated that he thought it was possible "for the big oleraceous things we know as *Cucurbita pepo* to have developed from a plant much like *C. texana.*"

In the next decade, A. T. Erwin, another botanist who devoted much time to the study of cucurbits, also observed the Texas gourd in nature and decided that it was an indigenous plant and not an escape. He stated that, except in minor characteristics, it agrees with the ornamental gourds (*C. pepo* var. *ovifera*) and clearly belongs to the same variety. He concluded that the preponderance of evidence indicated that it was the prototype of the cultivated forms of *C. pepo*.

The most recent treatment of *Cucurbita texana* is in the 1970 *Manual of the Vascular Plants of Texas,* by Donovan Correll and Marshall Johnston, in which it is listed as an endemic, rare but abundant where found, occurring "in debris and piles of driftwood, often climbing into trees, along several rivers, especially the Guadalupe, that drain the Edwards Plateau" in central Texas.

Is *Cucurbita texana* a truly wild plant, or is it naturalized? Unfortunately, I know of no way by which this question can be answered with certainty. Clearly the botanists who are most familiar with it regard it as a wild plant, indigenous to Texas. Naturalized gourds not dissimilar to *C. tex-*

ana have been reported from Illinois, Missouri, and Alabama. These are generally assumed to represent escapes from cultivation of ornamental gourds, which, if true, would indicate that indeed at least one form of *C. pepo* can revert to a feral existence. We do not know, however, whether ornamental gourds were grown in the area where *C. texana* occurs before 1848, at which time this plant first became known to science.

That the so-called naturalized gourds in other states might have come from Texas seems rather unlikely, but it is clear that the Texas gourd can be grown north of Texas, for it matured fruit for Bailey in New York, for Erwin in Iowa, and for me in Indiana.

In spite of the fact that the Texas gourd has been called a separate species, it is quite clear that it is nothing more than a variety of *Cucurbita pepo,* and several botanists in the last century treated it as belonging to *C. pepo.* Recent authors, however, have generally continued to treat it as a species, perhaps following Bailey, who in 1943 wrote: "In view of the fact that the connection between *C. ovifera* [= *C. pepo* var. *ovifera*] and *C. texana* has not been demonstrated in practice, I prefer in my time to keep the two things separate in nomenclature, that we may easier analyze our discussions." The connection between the two has, however, since been demonstrated. In a study of many characteristics of various species of *Cucurbita* using a numerical analysis, it has been shown that the two are very similar, and hybrids between the two are fertile. Moreover, quite recently Hugh Wilson and Thomas Andreas have produced biochemical evidence that indicates that *C. texana* is best regarded as belonging to the same species as *C. pepo.*

Thus we have a candidate for a progenitor of *Cucurbita pepo* north of Mexico in the Texas gourd, although this by itself does not necessarily lead to the conclusion that *C. pepo* had an independent origin north of Mexico. On the other hand, the problem of explaining why *C. pepo* came to eastern North America considerably earlier than corn and most other Mexican plants is eliminated if *C. pepo* had an independent origin north of Mexico.

Why did people ever domesticate a wild gourd in the first place? Certainly its rather fragile rind makes it inferior to the bottle gourd for the ordinary uses of a gourd. There is little in the literature to suggest that *Cucurbita pepo* had extensive use other than as food, although its use as a utensil is recorded in a few places. It could perhaps also have served as a rattle at times. What little flesh the Texas gourd has is extremely bitter. Therefore, we would have to assume that if it were used as a food plant, it was the seeds that were eaten, as may well be true of the wild gourds that gave rise to the other domesticated species of *Cucurbita*. The plants could have originally been cultivated for the seeds, as was pointed out above.

If we accept an independent domestication of *Cucurbita pepo* north of Mexico from *C. texana*, how is the much earlier domestication of *C. pepo* in Mexico to be explained? Although in the past other species have been pointed to as the possible progenitor of *C. pepo*, today it is recognized that the only possible progenitor would have to have been *C. texana*, assuming, of course, that it is a truly wild plant. Did early Mexicans carry the seeds of it southward, where it also underwent domestication? This seems rather unlikely, and it is more probable that *C. texana* or a plant very similar to it had a far more extensive distribution in earlier times than it does today. In fact, it could well have grown as far north as Illinois, which could account for the gourd in the early archaeological record. How then are we to account for its extinction everywhere but in Texas? This is a difficult question to answer, and there is no clear answer, but one might suggest as I have done for the bottle gourd that the wild plant may have grown in the very sort of habitats that people chose for their early settlements, and that humans eliminated it as a wild plant throughout most of its range by excessive harvesting or destruction of its habitat, or that it was hybridized out of existence by the cultivated form. But of course this apparently did not happen in central Texas.

If *Cucurbita pepo* were the only plant found in the archaeological deposits in Missouri dating to about 2000 B.C., one might build a strong case for the independent origin of agriculture north of Mexico, but it is not. The bottle gourd (*Lagenaria siceraria*) is found in the same levels as *C. pepo* in Missouri, and it most likely had to come from Mexico, which seemingly would considerably weaken the case for a separate origin of agriculture in eastern North America. Nevertheless, some examination of the bottle gourd is in order, for it would be enlightening to know if it came north from Mexico as a domesticated plant.

There are many enigmas in the history of the bottle gourd. It is not definitely known anywhere in the world today as a truly wild plant, but it is generally assumed to be native to Africa. The archaeological record reveals that it was in Mexico by 7000 B.C., far earlier than it is known from Africa. How it could have reached America from Africa by that early date is not known. Some have postulated that it must have been carried by people, whereas others have maintained that a fruit drifted across the Atlantic. Since bottle gourds have been shown to float for more than a year in ocean water and still have viable seeds, the latter explanation appeals to me. Rivers could have carried the bottle gourd to the ocean, then the ocean currents most likely would have carried the gourds to the coast of Brazil, and from there it could have spread over the Americas as a trade item, if not naturally as a camp-following weed.[3]

To carry the speculation even further, we might postulate that if a bottle gourd made a voyage across the Atlantic, one also might have floated to the coast of southeastern

[3] An interesting account of water serving as the means of dispersal for gourds comes from the travels of Cabeza de Vaca, called to my attention by Dee Ann Story, an archaeologist at the University of Texas. From Bandelier's account of Cabeza de Vaca we learn:

In the afternoon we crossed a big river, the water being more than waist-deep. It may have been as wide as the one of Sevilla, and had a swift current. At sunset we reached a hundred Indian huts and, as we approached, the people came

North America, not necessarily from Africa but perhaps from Mexico. The beaches of Florida as well as other parts of the Atlantic and Gulf coasts of the United States regularly receive tropical fruits and seeds carried by ocean currents. Bottle gourds could have reached the Gulf of Mexico by floating down rivers of eastern Mexico. Thus, people of the eastern United States could have encountered gourds with no knowledge of their cultivation. That a bottle gourd came to the United States in such a manner is not impossible, but, I must admit, it seems more likely that it came north from Mexico overland.

But did it arrive as a domesticated plant? Although the bottle gourd was in Mexico by 7000 B.C., we do not know when it became an intentionally cultivated plant. It is impossible to say when it became domesticated, for there are no wild bottle gourds with which to compare the archaeological material. The earliest bottle gourds known in the United States have rather small seeds, which might be expected in a wild form, so the possibility exists that it was not domesticated. Maybe it came north as a weed that followed human beings in their travels. In the final analysis, however, I will have to admit that the evidence perhaps favors agriculture coming to eastern North America with *Cucurbita pepo* and the bottle gourd, but the final word is

---

out to receive us, shouting frightfully, and slapping their thighs. They carried perforated gourds filled with pebbles, which are ceremonial objects of great importance. They only use them at dances, or as medicine, to cure, and nobody dares touch them but themselves. They claim that those gourds have healing virtues, and that they come from Heaven, not being found in that country; nor do they know where they come from, except that the rivers carry them down when they rise and overflow the land.

Through recent attempts to retrace the route of Cabeza de Vaca, it is possible to identify the "big river" as the Rio Grande, and the crossing was probably near what is now Roma, Texas. The problem of identifying the gourd is more difficult. It might have been a species of *Cucurbita*, but I think that the bottle gourd is more likely, for it is commonly used by Indians to make rattles. If it were the latter, it most likely would have come from cultivated fields unless we assume that in 1535 this species did grow as a wild plant or weed in the Southwest.

A Japanese wine or water gourd (*Lagenaria siceraria*).

yet to be written. The next archaeological dig might contribute to it.

It is hardly surprising to encounter the gourd at the earliest levels of agriculture in eastern North America, for it appears with the beginning of agriculture in many places. Although probably never more than of minor use as food, the bottle gourd early became almost indispensable as a water jug and container and used in many other ways, for example, as musical instruments and art objects. Little wonder, in view of its great importance to people, that a rich mythology grew up around the gourd. I devoted a chapter to this subject in *The Gourd Book*, but I had no idea how extensive the mythology was until Norman Girardot later sent me a chapter from his *Myth and Meaning in Early Taoism*, where in beautiful style he has shown the great significance of the gourd in China. He, quoting from Ho, tells us that humankind emerged from a gourd and escaped the flood by floating in a gourd. We also learn much more: for example, seed grains magically appeared in a gourd, and the gourd vine served as a ladder for people to ascend to the heavens. Other plants have served similar purposes in myths, but none perhaps to the same extent as the gourd. One is tempted to derive the origin of agriculture from early cultivation of the gourd; there is, however, one serious drawback. According to the archaeological record, the first appearance of agriculture was in the Near East nearly ten thousand years ago with the domestication of wheat and barley. So far as I am aware, there is no evidence for the bottle gourd in early times in the Near East, and it is absent from the early mythology of that region.

There is one other cucurbit that deserves discussion before passing on to other subjects—the genus *Luffa*. The fruits of one of its species are now widely sold in the United States for use as a cosmetic sponge. My interest in *Luffa* was initiated when I was writing *The Gourd Book*, for I found that the exact number of species in the genus was uncertain and that some species were native to the tropics of the Old World and one or two were native to tropical America. How many species are there, and how does one account for the presence of species on both sides of the ocean?

These questions prompted an investigation, and with the help of many people I was able to assemble seeds from a variety of places, which allowed me to grow the plants in the greenhouse for a detailed study. With Ed Schilling, a former student of mine now at the University of Tennessee, I concluded that there were five or six species in the genus. Since most of these species do not have common names in English, it will be necessary to use their scientific names. Four of the species are Old World in origin: *Luffa acutangula*, which has a wild variety with extremely bitter fruits and a domesticated variety with nonbitter fruits widely used for food in southeastern Asia; *L. aegyptiaca* (often given as *L. cylindrica*), which also has wild and domesticated varieties, the latter known as the sponge gourd, dishrag gourd, or simply loofah and the one mentioned above, used as a cosmetic sponge as well as in other ways; and *L. echinata* and *L. graveolens*, both wild species. In the New World there is at least one wild species, and we shall use the name *L. operculata* to refer to it. It seems likely, for reasons that I shall not go into here, that the genus had its origin in Asia, so the problem is to explain how it got to the Americas.

First of all, the genus might have had its origin on Gondwanaland, the great southern land mass that existed in early geologic times and that later broke apart to give rise to the southern continents and India. If the genus existed on Gondwanaland, possibly a part of it had reached the area that was later to become South America. I regard this as rather unlikely, for it seems improbable that the cucurbit family had come into being at the time when Gondwanaland was in existence. In fact, probably at the time of the breakup of Gondwanaland, few or none of the flowering plants had yet arrived on the scene.

A second possibility is that people are responsible for carrying it across the ocean. The American species, *Luffa operculata,* is used today in folk medicine, chiefly for the treatment of sinus trouble,[4] but apparently it is not cultivated for that purpose, wild plants being used. Some of the Old World species are also reported to have medicinal use, so one might suppose that people took it with them to America because it was a useful plant. When the plant arrived in the Americas, it found a suitable habitat, evolved into a new species (or possibly two) quite different from any of the Old World species, and eventually attained an extensive distribution all the way from Mexico to Brazil, including islands off the coast of Brazil and Ecuador. One

[4]From personal experience I can say that they have an effect. In removing the seeds from the dry fruits of *Luffa operculata,* I noticed that a fine dust was present. To remove all of the seeds from the network of fibers takes several minutes and I must have breathed some of the dust. Later in the day I noticed that my nasal passages were quite dry, and in the evening I had a headache. A year later I did the same thing with the same results. But it hasn't happened a third time, for I now wear a dust mask when extracting the seeds.

Other uses for the wild *Luffas* are ornamental in nature. We have found that the fruits of *Luffa operculata* make interesting Christmas tree ornaments, either painted or unpainted. The fruits can also be used to advantage in dried winter bouquets.

Fruits of wild species of *Luffa:* (*top left*) *L. acutangula;* (*top right*) *L. aegyptiaca;* (*center*)
L. operculata, with outer part removed from fruit on right to show network of fibers;
(*bottom left*) *L. graveolens;* (*bottom right*) *L. echinata.*

can only speculate how long it took for the new species to
evolve and attain its present wide distribution. However,
if people made such a voyage at 5000 B.C., about as early
as such voyages have been postulated, there would hardly
have been time for that to occur. Therefore, I don't think
this hypothesis can have much support.

This leaves only one possibility. Seeds or fruits managed, without human aid, to cross the ocean—and present
distributions suggest that it was the Pacific—after the continents had more or less assumed their present positions.
Botanists postulate that long-distance dispersal of plants
has occurred by the agency of birds, wind, or water. There
is no evidence that birds eat the seeds, and the seeds are

Luffa operculata used in winter bouquet (arrangement by Laura Soria).

too heavy to be carried by air, so this leaves water as the most likely agent. Indeed, a water dispersal of *Luffa* was suggested by the observations of the naturalist H. B. Guppy in 1906. He had seen fruits of the wild variety of *Luffa aegyptiaca* floating down streams in Fiji. So he decided to see how long fruits and seeds would float. He found that the fruits would not float more than a week, but that sixty out of one hundred seeds were still afloat after two months. So he concluded that seeds of the variety may at times be dispersed by ocean currents. Such would account for the distribution of the variety on islands in the South Pacific. Two months, however, would hardly be enough time for the seeds to float to the Americas.

I decided to carry out some tests of my own. Twenty-five seeds and three fruits of each of the wild species or varieties were placed in bowls of seawater, and as the water evaporated, distilled water was added to keep the salt con-

centration as constant as possible. Most of the seeds sank
almost immediately. Nine seeds of *Luffa aegyptiaca* managed to float for over one month, and when planted, three
of those seeds germinated. Fruits of two of the species
floated less than two months, but those of the others did
considerably better. One of the fruits of *L. graveolens* floated
for eighty-nine days, and two out of its twelve seeds were
viable. One fruit of *L. operculata* remained afloat for two
hundred days, and later two of its fourteen seeds germinated. The record was achieved by *L. acutangula,* for two
fruits floated for one year, at which time they were removed
and the seeds were planted. Twelve of the sixty-one seeds
germinated.

No far-reaching conclusions should be attempted from
this simple experiment. The ability of seeds to float in a
bowl of salt water is hardly an adequate test for their ability to survive in the ocean. Moreover, it should be pointed
out that only a small number of fruits was tested, and for
each species only one accession was represented. It is possible that there is variability within a species for the ability of its fruits to float. It does seem likely, however, that
island hopping by drifting fruits is a distinct possibility
for some of the species, and one might postulate that fruits
of *Luffa acutangula* might travel considerable distances. It,
however, is not a species particularly closely related to
L. operculata. The two hundred days that a fruit of *L. operculata* floated is probably not sufficient time to account
for a voyage across the Pacific, but perhaps the species that
later evolved into *L. operculata* had fruits capable of staying
afloat for a longer period. One might even postulate that
a fruit or seeds drifted across on a log, perhaps not too
unlikely, for the vines of *Luffa* often climb trees, and some
species occur on the coast. Although proof has not been
presented that American *Luffa* came by floating, that is,
I feel, the best hypothesis.

From the material of *Luffa* that was grown in the greenhouse, we made a detailed study of the characters of the
flowers, roots, and leaves and the flavonoid pigments in

the plants. The measurements were thrown into a computer, which made the appropriate correlations, and as a result we can say that the American *Luffa* is most similar to *Luffa graveolens*, which was not terribly surprising, for we had already come to that conclusion on the basis of inspecting the various species in the greenhouse. *L. graveolens* today is found in India and northern Australia, and on the basis of its similarity to *L. operculata* I am inclined to believe that it, or more likely the progenitor of it and *L. operculata*, was the species that reached America to give rise to *L. operculata*.

In addition to morphological comparisons, there are other ways to study the relationships of species. One that we had used with other genera was the hybridization of species. Members of the same species will generally cross with ease and produce fertile offspring. Crosses between species considered to be closely related usually are made fairly readily, and the seeds give rise to vigorous hybrids that show some fertility. Distantly related species will not cross at all, or if they do, the hybrids are usually sterile or show highly reduced fertility. My use of "usually" indicates that exceptions are known, but in many groups of plants—for example, the annual sunflowers—the fertility of the hybrid agrees with the relationship as judged on morphological grounds, the most similar species producing hybrids showing greater fertility than the hybrids of the rather dissimilar species.

Some of the hybrids of *Luffa* did not follow this pattern. It took more than 150 attempts before a single seed was secured in crosses of *Luffa operculata* and *L. graveolens*, yet each of them will cross fairly readily with the other species of the genus. These crosses with other species usually produce normal flowers, but they are sterile or nearly so. The one hybrid secured between *L. operculata* and *L. graveolens* produced numerous buds, but they seldom opened, and when they did open, the flower parts were all quite abnormal. Even more surprising was the failure to secure seeds in crosses between Ecuadoran and Brazilian accessions of

L. operculata, and this is the reason for thinking that we may be dealing with two species in spite of the great morphological similarity.

Are we to say that the relationships based on morphology are incorrect as concerns *Luffa operculata* and *L. graveolens?* I hardly think so, for the morphological similarity is due to a large number of genes, whereas the failure to cross could result from a single or only a few genes. It could even be advantageous for closely related species to be unable to cross with each other if they are growing together, for all seeds producing hybrids would be a loss for the perpetuation of the species. However, this explanation hardly holds for *L. graveolens* and *L. operculata,* for they are an ocean apart.

CHAPTER 2

The Totora and Thor

In 1973 I received a letter from Robert W. Long, who was then editor of the *Plant Science Bulletin,* asking me if I would write an article on some phase of my ethnobotanical investigations. I couldn't think of anything to write about, and as I didn't want to do a rehash of something that I had already published, I was about to decline. Both my wife and the secretary of the botany department, Virginia Flack, told me in no uncertain terms that I couldn't do that, for Bob was a former student of mine and a good friend. So I decided to give the matter more thought.

A few evenings later I was showing slides of Ecuador to some students, and one of the slides was of a highland lake around which a large number of totora reeds were growing. I didn't know a great deal about this plant; in fact, I couldn't even recall its name at the time, but as I saw it on the screen I vaguely recalled that it was the same plant that Thor Heyerdahl claimed had been introduced from Peru to Easter Island by ancient Peruvians. The next day I looked up Heyerdahl's *Aku Aku* (1958), a popular account of his trip to Easter Island, and I found that he had indeed made such a claim for the totora. Was the plant on Easter Island really the same as the American plant? I wondered, and I realized that I might have a subject for an article for Bob.

Thor Heyerdahl, born in Norway in 1914, came to the attention of the world in 1947 when he sailed the *Kon Tiki,* a primitive vessel made of balsa logs,[1] from South America

---

[1] Not to be confused with the balsa rafts of Lake Titicaca, which are made from a reed. Balsa logs come from a tree, *Ochroma lagopus.*

far into the Pacific. The voyage was undertaken to prove that it would have been possible for people to have made long ocean voyages in prehistoric times. Moreover, Heyerdahl felt that certain Pacific Islands had been colonized from the Americas.

Heyerdahl belongs to the school of "diffusionists," who believe that most, if not all, inventions are made only once and then diffuse to other parts of the world. For example, the diffusionists believe that the many similarities between the prehistoric cultures of the Old World and the New World result from the introduction of traits or ideas from the Old World to the New. The "independent inventionists," on the other hand, hold that people living in similar environments in different parts of the world will come up with similar inventions when faced with similar problems. Moreover, the independent inventionists generally believe that long ocean voyages were rare, if they occurred at all, before the time of Columbus. The trip of the *Kon Tiki,* and Heyerdahl's later voyage across the Atlantic in a reed boat made in Africa,[2] proved that such voyages would have been possible. Heyerdahl's voyages have forced many to reevaluate their conclusions regarding prehistoric travels. Probably most scientists today would accept that there were voyages from the Old World to the New World long before Columbus, although probably very few of them would agree that the cultures in the Americas were strongly influenced by such prehistoric contracts.

I have great admiration for Heyerdahl as a seaman, explorer, and adventurer, but I have some reservations about his scientific work, chiefly because of the way in which he used, or misused, plant evidence to support his claims. He was not alone in this, for some cultural geographers have drawn similar conclusions. The one plant that is generally thought to have been carried from the Americas to Pacific islands in prehistoric times is the sweet potato. It seems clearly to be of American origin, and it

[2] This reed was papyrus and is not the same plant as the totora.

has generally been accepted that it was in Polynesia before the Spanish began their great ocean voyages of the sixteenth century, although this idea is disputed by Brand. It seems likely that roots of the sweet potato would have served as the material for propagation, and these would have to have been introduced by humans. Since the sweet potato does on occasion produce seed, some writers have suggested that the seeds might have been carried by birds or currents or on drifting logs to the islands, which, if true, would mean that people would not have been involved. Even if humans were responsible for the sweet potato's introduction, they would not necessarily have had to be ancient Peruvians, for Polynesians might have made the journey to South America and carried roots back home with them.

Easter Island is quite small, about fifteen miles long and eleven miles wide, and is located two thousand miles west of Chile. Its historic discovery was by the Dutch Admiral Jacob Roggeveen on Easter day in 1722. The island has been of considerable interest because of its monolithic stone heads and tablets with hieroglyphic characters. How the statues were made and erected is now understood, partly through Heyerdahl's own work, but insofar as I am aware, the hieroglyphic script has yet to be deciphered. When and by whom the island was settled has also been a subject of interest. Heyerdahl claims that some of its people came from America, but few others who have investigated the subject would agree, holding that all of its early settlers are Polynesians in spite of some of the differences between the island's culture and that of other parts of Polynesia. The history of the island since 1722 has not been a particularly pleasant one, but I shall not go into that. In fact, my object is not to examine the people but only the plant known as the totora, which Heyerdahl calls upon to support his contention that there was a prehistoric American connection.

Heyerdahl claims that the totora on the island was the same as the American plant that, according to him, was cultivated under irrigation all along the coast of Peru in prehistoric times. The plant was used to make mats and boats on Easter Island as was done in Peru. Heyerdahl states that the pollen record and archaeological evidence show that it was on the island before the first Europeans arrived, but he says that man must have brought it to the island, for it generates "only by new shoots from the suckers," and thus it could not have been carried to the island by the wind, the ocean, or birds. Moreover, he points out that birds are very rare on the island. Finally, he claims support from the botanist Carl Skottsberg, who maintains that the species was the same as the one in Peru, and furthermore that human introduction would be the easiest way to explain the difficult problem of how the plant got to Easter Island.

Heyerdahl is quite correct insofar as I can determine in several of his conclusions. The totora was used in much the same way on Easter Island as it was in Peru. This, of course, he uses to imply that humans introduced the plant and the ways of using it at the same time. However, could not the people, who perhaps found the plant already on the island, have independently discovered how the plant could be used? After all, the ways in which the plant is used—for mats and boats—are not terribly complicated, as will be shown later. Making this statement may label me as an independent inventionist, so I hasten to add that the ways in which the plant is used, if not independently invented, need not have come from the Americas. Reeds related to the totora were used in similar ways in the Old World.

There seems to be no doubt that the plant was on the island when Europeans first arrived. The pollen record might be helpful in determining when the plant first arrived. Unfortunately, O. H. Snelling, who was to publish on the pollen record, has never done so.

That there were only a few birds on the island in historical times is mentioned by others, so there is no reason to question that statement. It does not necessarily eliminate birds as a possible agent of dispersal, however, for it would take only one. Moreover, one wonders if there may not have been a much denser cover of vegetation when the first people arrived and that consequently birds were more numerous at one time.

Heyerdahl gives the scientific name of the totora as *Scirpus riparius,* which is the name formerly used for the plant now known as *S. californicus.* This reed or bulrush, a member of the sedge family, has an extensive range from California to Mexico, throughout parts of Central America, and over much of South America from Colombia to Chile, in addition to Easter Island and Hawaii. In Peru two races are recognized, one found in Lake Titicaca and a few other places in the highlands and a second found both in the highlands and on the coast. It would be the latter that Heyerdahl postulated was carried to Easter Island. In investigating the nomenclature, however, I learned that at one time the plant on Easter Island had been regarded as a distinct species, but in the recent treatment of *Scirpus* by an authority Tetsuo Koyama, it is regarded as a variety (var. *paschalis,* Latin for Easter). Koyama also reports this variety from Chile, so the plant could still have gone to Easter Island from South America. (There can be but little doubt that the plant went from the Americas to Easter Island and not in the reverse direction, for all its close relatives are American.) Thus, my original question whether the plant on Easter Island was the same species as the American plant was answered. Whether the Easter Island plant was a distinct variety from the one in Peru perhaps deserved further study.

The next question I addressed before starting my article was the mode of reproduction of the plant. It will be recalled that Heyerdahl had stated that the plant reproduced only by "suckers." This would mean vegetative reproduction and imply that it did not reproduce by seed. At the time, I was not in a position to do more than search the literature. I learned that at times new plantings of totora are made by transplanting portions of the rootstock (or

rhizome), which is probably what Heyerdahl meant by
"suckers." Several descriptions of the plant, however, referred to the achenes. Achenes are one-seeded fruits, and
nonbotanists would probably call them seeds as they do
grains of wheat or corn, which also are fruits. I did not find
any information about whether the seeds in the achenes
would germinate, but I could see no reason why they would
not. If they would germinate, it destroyed a large part of
Heyerdahl's argument, for immediately the possibility that
birds introduced the achenes to Easter Island came to mind.
Of course people also could have introduced achenes.

   Finally, I tried to find evidence for the prehistoric cultivation of the totora on the coast of Peru. Although I
did find records of its present-day cultivation, I could find
nothing in the archaeological record or in the accounts of
early visitors. From the latter, however, I did learn a lot
about the plant and its early use in the Americas.

   The most extensive account of the plant is by Bernabé
Cobo in his *Historia del Nuevo Mundo*. Cobo came to the
Americas in 1596 and spent fifty years traveling to many
places. His manuscript was completed in 1653 but not published until 1890. I quote from his chapter entitled, "Of
the Grasses That Are Found in the Indies That Are the
Same Kind as Those of Spain," using my own translation.
After stating that he sees plants in the New World like those
in Spain but that they are all wild, he goes on to write:

   The plant and grass that is most commonly found in all parts
   of this land [the New World] is the rush, which usually grows
   on the banks of the lakes and rivers in marshes and swamps;
   there are many different ones, all of which this land produces
   abundantly. The first and largest kind of rush is the *enea:* called
   Tutura by the Indians of Pero, *Totora* by the Spanish, and *Tule*
   by the Mexicans. In the provinces of Collao where they grow
   in great abundance on the banks of the great lake Chuciuto [Lake
   Titicaca], particularly certain of them that are triangular; they
   serve as pasture for the beasts, and the roots, which are white
   and tender, serve as food for the Collas Indians; that whose
   roots is called *Cauri,* and many bundles of them are sold in the

Esteras for sale at the market in Quito. The baskets are made of *carrizo*, a grass. (The photographs in this chapter are reprinted by permission of *Economic Botany* 32 (1978): 222–36; copyright © 1979, Charles B. Heiser and the New York Botanical Garden.)

plazas of the Pueblos for this purpose as are other sustenances. From the dry rushes the Indians of Peru make mats (*esteras*) and boats (*balsas*), not only for crossing the rivers but also for boarding and fishing in the sea, especially from *Totora*, the name given to the thick triangular rush, or *enea*. . . .

In some places the Indians make admirable mats from the narrow and long rushes, especially in the city of Lima and in the pueblo of Lambayeque, diocese of Trujillo. Here they are called *Petates*, which is the Mexican name, and from them they make all sizes like carpets and they serve as such for the pedestals of the altars, in the drawing rooms of the women, and they are accustomed to putting them under the expensive rugs, and the travelers cover their traveling bags for sleeping and other baggage with them, because they protect them from the rain.

The observations of José de Acosta are also of interest. Padre Acosta came to the New World as a Jesuit missionary in 1571. Most of his time was spent in Peru, but he also

House made of totora on an island in Lake Titicaca. The island itself
is composed largely of totora reeds. Bundles of harvested reeds may be
seen in the background.

traveled widely to other parts of the Americas before
returning to Spain in 1587. His *Historia Natural y Moral de
las Indias* appeared in 1590, and I quote from an early
English translation:

It [Lake Titicaca] brings forth a great aboundance of reedes,
which the Indians call *totora,* which serves them to a thousand
vses; for it is meate for swine, for horses, and for men; they make
houses therewith, fire, and barkes. To conclude, the Uros in this
their *totora* finde all they have neede of. . . . There are whole
villages of these Uros inhabiting in the Lake in their boates of
tortora [*sic*]. . . . The bridges they made were of reedes plaited,
which they tied to the bankes with great stakes, for that they
could not make any bridges of stone or wood. . . . [on the] river
draining the great lake Chucuito [Titicaca]. . . . they did binde
together certaine bundles of reedes, and weedes, which do grow
in the lake that they call Totora, and being a light matter that
sinkes not in the water, they cast it vpon a great quantity of

reedes; then, having tied those bundles of weedes to either side of the river, both men and beasts goe over it with ease. Pasing over this bridge I wondered, that of so common and easie a thing, they had made a bridge, better, and more assured than the bridge of boates from Seville to Triana. I have measured the length of this bridge, and, as I remember, it was above three hundred foote.... There is another fishing which the Indians do commonly vse in the sea, the which, although it be lesse, yet is it worthy the report. They make as it were faggots of bul-rushes or drie sedges well bound together, which they call Balsas; having carried them vppon their shoulders to the sea, they cast them in, and presently leape vppon them. Being so set, they lanch out into the deepe, rowing vp and downe with small reedes of eyther side, they goe a league or two into the sea to fish, carrying with them their cordes and nettes vppon these faggots, and beare themselves thereon.... Truely it was delightfull to see them fish at Callao of Lima, for that they were many in number, and every one set on horsebacke, cutting the waves of the sea, which in their place of fishing are great and furious, resembling the Tritons or Neptunes, which they paint vppon the water, and beeing come to land they drawe their barke out of the water vpon their backes, the which they presently vndoe, and lay abroade on the shoare to drie....

More is learned from Garcilaso de la Vega, who was the son of an Inca princess and a Spanish nobleman. Thus, from his mother's side of the family he had information about the Indians that few Spaniards could obtain. However, since his observations were not put into writing until 1600, some forty years after he had left Peru for Spain, perhaps we must allow for certain lapses of memory. He gives an even longer account of the "bridges" made of sedges near Lake Titicaca than does Acosta. After describing the balsas or boats made of balsa logs, he goes on to say that the Indians "make another kind of boat of rolled up bundles of rushes of the thickness of a bullock; they fasten them securely together; from the centre towards the bow the size diminishes, like the bow of a ship, so that they may cut through the water. The top of the bundle is flat for receiving the cargo." This description fits fairly well the balsas used on Lake Titicaca today, but the method of propelling them which he goes on to describe is quite different. Two pages further on, he discusses fishing in the ocean and tells that the fishermen, when they navigate the sea, sit on their knees upon the bundles of reeds, which is not unlike what they do presently at Huanchaco, Peru.

Thus from three different writers we have accounts of the use of totoras, but not one of them says anything about the reeds being cultivated. Cobo, in fact, says that they are not of the "garden."

At last I had the material for an article for the *Plant Science Bulletin.* The plants on Easter Island and in Peru were indeed the same—the totora was used in similar ways in the two areas—but I could find no evidence that it had been cultivated on the coast of Peru as Heyerdahl claimed. Furthermore, the plant did apparently produce seed, so although I could not deny the possibility that it might have been introduced by humans, I concluded that an introduction by birds was far more likely.

Easter Island and many other oceanic islands were covered with vegetation long before the islands were inhabited, so it is obvious that the seeds or other propagules of plants can travel great distances in or over water. They manage to do this by being transported by air, water, or birds. Extremely small seeds or spores of ferns and other non-flowering plants can be carried extremely long distances by the wind. The achenes of the totora are quite small, less than one-eighth inch long, but it is probably unlikely that they would be carried any great distance by the wind. Some plants have seeds or fruits that will float for long periods of time and remain viable, so ocean currents may be responsible for their dispersal over long distances. Seeds or other propagules might also be carried by floating logs. Tests have not been conducted, but I feel it unlikely that ocean currents were involved in the dispersal of the totora.

So this leaves birds as the possible agent for dispersal. Botanists have postulated that fruits of sedges are often dispersed by birds, and the same species of some sedges not used by people are also found on both continents and oceanic islands. There are various ways in which birds might carry the achenes. First, the achenes might be eaten and then passed undigested in a new area. An anthropologist friend of mine, Jerry Epstein, has suggested, however, that it would take a highly constipated bird to carry a seed all the way from some place in the Americas to Easter Island. Second, a bird feeding in sedges might get achenes caught in its feathers, and they might remain there for a considerable period of time. Finally, birds walking in marshy areas might get mud containing achenes lodged on their feet. Which one of these ways, if any, would explain the distant dispersal of the totora is not known. Sherwin Carlquist, who has made a special study of the introduction of plants to islands, estimates that 70 percent of the seed plants on Easter Island arrived with birds.

It was pointed out earlier that the totora also occurs in Hawaii, but very little is known about the plant there. It has been used for making mats in Hawaii, but apparently it is not terribly important for that use, perhaps because several other plants are available. Questions arise. Did the plant come to Hawaii from Easter Island, or, for that matter, could Hawaii have been the source of the plants on Easter Island? These questions cannot be answered at present.

After all my reading I still had other questions about the plant. Certainly it would be desirable to know to what extent, if any, the plant reproduced by seed. How widely was it cultivated? Could it be considered a domesticated plant? And just how similar were the plants of Easter Island to those of coastal Peru? Although I had seen the plant in highland Ecuador and at Lake Titicaca on previous trips, I had never given it more than a casual inspection. One way to learn more about the plant would be to grow it in the greenhouse. I began a letter-writing campaign which brought me helpful information about the plant from South America, but I failed to receive any achenes.

Much to my surprise, I did obtain some from Easter Island
with help from Bob Meier, an anthropology professor at
Indiana University who had spent some time on the island.
He suggested that I write to John Tuki, and a few months
later I had achenes. The seeds germinated readily, and the
plants proved surprisingly easy to grow to maturity in the
greenhouse. So it was clear that the Easter Island plant
produced viable seed, but I still needed material from
South America. I finally decided the easiest way to get it
was to go myself, so I decided to go there on my sabbatical
leave the following year. It was a decision I was happy
with, for there is no place that I would rather visit than
the Andes. The trip, of course, would also allow me to
make observations on other plants that I had under study.
Early in 1975 my wife and I went to Ecuador. After getting
settled in a hotel, I got in touch with friends for their
advice on our search for totoras.

Before going into the details of our travels in Ecuador
and Peru, a few words on the name of the totora are in
order. Throughout this chapter I am using the name totora
for *Scirpus californicus*. In western South America, totora,
sometimes spelled slightly differently, as, for example,
tutura, is the name most commonly employed, whereas in
Mexico and North America, the plant is usually called *tul*
or *tule*. Other Indian names such as *matara* and *mirme* are
sometimes used in South America, and in Spanish the plant
is sometimes referred to as *junco, enea,* or *juncia.* Unfortunately, as is often true for common names of plants, these
same names may be employed for other plants, some of
which, such as the cattail (*Typha*) and the rush (*Juncus*), do
not even belong to the same family as the totora (*Scirpus*).
Often after arriving at the place where I had been directed
in my search for totoras, I would find cattails or rushes.
This was not surprising, for all these plants grow in wet
places and have more or less the same aspect. There is
also a resemblance of many of these plants to grasses, and
some people, such as Cobo, use the word as an inclusive
term for all such plants. A number of places, both in Ecua-

dor and Peru, are called Totora or Totoral (a place where
totoras grow). In visiting some of the places, such as Totoras near Ambato, Ecuador, I found totoras growing, but in
others, such as Totoracocha near Cuenca, Ecuador, I found
neither totoras nor a lake (*cocha*), and the area was occupied by houses.

Our first trip was to Otavalo, about one and one-half
hours by car north of Quito. This small, attractive town is
at 8,500 feet and is the site of the market of the Otavalo
Indians, widely known for the fine ponchos and other
woven goods they make. The lake, San Pablo, where the
totoras grow, is only a short distance from the town, and
I soon found myself examining the plants for achenes. As
I was doing so, an Otavaleño came up and asked me what
I was looking for. I explained that I was trying to learn
if the totora produced seeds. He replied that it did and
then proceeded to take some dried flowers to examine
himself. After a moment's search, he asked me, *"Tiene
usted un lente?"* (Do you have a lens?). That he would know
what a lens was came as something of a surprise. I handed
him my lens, and he put it to his eye to examine the flowers.
We didn't find many achenes, but I took the occasion to
ask him about the boats (here called *caballetes,* diminutive
of the Spanish, *caballo,* for horse), for I didn't see any on
the lake. He said that he had one and asked me to come
to his house to examine it. There I saw it and some mats
of totoras that the family made for sale and that they also
used to cover the dirt floor of their one-room house. The
man and his wife and mother were happy to pose for a
photograph, and he insisted that his dog be included
as well.

The *caballetes,* which are used for fishing or hunting
ducks in the lake, are always carried to the home after
their use, perhaps so they won't be stolen but probably
more because they become waterlogged if left in the water
for long periods of time. The boats of the Otavalo region
are rather crudely constructed in contrast to those that I
would later see in Peru. They were hardly more than a

bundle of reeds. How ancient the tradition of using totora craft is in northern Ecuador is unknown.

The mats made from the totora are more important than are the boats. These mats serve the Indians in many ways today just as they did in ancient times, being used for beds,[3] floor coverings, room dividers, fences, windbreaks, and the like. The manufacture of mats has become a cottage industry at Lago San Pablo and in many other places. Quito is one of the main outlets for the mats made at Lago San Pablo. A family of three can make three large mats (about four by six feet) in an eight-hour day, which does not include the time for cutting the totoras. After the totoras are cut, they are allowed to dry, then they are moistened and rolled with a stone to flatten them, cut to the desired length, and woven into the mats. This method seems to be the common one not only in the Andes but in Mexico as well. At the time we were in Ecuador in 1975, the mats sold for about fifteen *sucres* (seventy cents U.S.) in Otavalo, and after being taken by truck to Quito the price was about twenty *sucres*. This may not seem like a lot of money for so much work, but it meant a great deal to an Indian family. In Quito the mats sell well and are found in the homes of both the rich and the poor. In Ecuador these mats are nearly always called *esteras,* the Spanish word for mats. The word *petate,* however, is recognized in southern Ecuador and proved to be in wider usage in Peru than *esteras.* As Cobo has told us, *petate* is the Mexican word for mat. Whether that word was introduced into South America in prehistoric times or whether it came in with the early Spanish from Mexico is not known.

Totoras are also made into fans, *aventadores,* and small baskets. The *aventadores* are almost indispensable in the highlands, where they are used to fan fires to get them started and to keep them going, a necessity at that altitude because of the rarefied air.

---

[3] In Peru there is a saying, *nacido sobre un patate* ("born on a mat"), implying a humble origin.

Otavalo Indians with *esteras* at Lake San Pablo, Ecuador. Their *caballete* may be seen behind the woman in the middle. The man is the one who helped the author search for seeds.

Before our afternoon was over at Lago San Pablo, I dug rhizomes of two totoras. Upon our return to Quito, they were mailed to Indiana University by way of Miami, where they were inspected for disease by the U.S. Department of Agriculture. My assistant, Lewis Johnson, planted them in the greenhouse immediately upon their arrival at Bloomington. Of the sixteen shipments from various places, eleven survived the journey. The rhizomes were insurance in case the seeds that I had collected did not germinate.

From Otavalo we went north by bus to Ibarra. When I visited the place thirteen years earlier, it had been a rather small, sleepy town, but now it was a bustling city. It held no particular attraction for us except for being near Yaguarcocha ("Blood Lake"), where I knew totoras grew in some abundance. This lake is of historical significance, for not too long before the Spanish arrived, the Inca had con-

quered the native Indians, and the lake was said to have
been red with the blood from the battle. I found recently
cut totoras drying around the edge of the lake, and I saw
a man and his wife carrying totoras to their house. The
people here were rather sullen,[4] quite unlike the Otavale-
ños. I found it difficult to engage them in conversation,
and they did not want their photographs taken. The most
important thing I learned there was that the totora set an
abundance of seeds. Yaguarcocha is at 7,200 feet, and later
I decided that the lower altitude there may have had something to do with the seed production. The taxi we had
arranged to take us back to Ibarra never did come, but
we were able to hitchhike our way back.

After spending a few days in Quito and making some
trips to lakes above 12,000 feet, where we failed to find
totoras, we set off with friends to Lago Colta, which is
near the city of Riobamba. In addition to collecting totoras,
I had another task to perform here. I had been in contact
with Eileen Maynard, an anthropologist who had worked
there a few years earlier, and she wanted me to take a gift
to her godchild. Having done so, we were received most
cordially in the area. The totoras grew extensively there,
and I knew from my reading that they were used for making mats and also for feeding livestock, good pasture being
in short supply. Certain people held rights to the totoras,
and sometimes others would raid their territories at night
to steal the plants, which attests to their importance there.
Some "cultivation" of the totora was also practiced; families who wanted to extend their holding of totoras would
sometimes transplant the rhizomes in low, wet areas near
the lake, where they would continue to grow. I examined
many plants, but I failed to find a single achene. Colta,
at over 10,500 feet, was higher than the lakes where I had
found seeds previously.

Although the stands of totora at Lago Colta appeared

[4]After several hundred years of oppression by the Europeans, they
have every reason to be so.

to be natural, I learned later that they were originally introduced by people. Sylvia Helen Forman, an anthropologist who had been at the lake a few years earlier, was told by an old man that the totora did not always grow at the lake. His father and some other villagers had seen it at Palmira, about thirty miles south of Colta, and they had brought some to Colta. The first planting was destroyed by ducks, but the second one was successful. That must have been about 1900. She also learned that they had once made boats of totora, and the man showed her how it was done, but later they brought in from the coast balsa logs, which they found made superior boats or rafts. Learning about the introduction of the totora at Colta, I wondered to what extent the totora had been introduced by Indians to other places where the stands appeared to be natural, as at Lago San Pablo and Yaguarcocha.

The last place in Ecuador where I collected totoras was at Llacao. I particularly wanted to see them, for Luis Cordero in 1911 had written that they were cultivated and cared for there. To get to Llacao we flew to Cuenca, which is in the southern highlands. Although today only a forty-five-minute flight from Quito, Cuenca until recently was rather isolated from the rest of the country and thus had preserved much of the colonial culture of earlier times. Although I had been in the city briefly on an earlier trip, this time I was able to explore it more fully. Among the things that we saw was the home of Cordero, which is marked with a plaque, for as well as being one of the country's early naturalists, Cordero had also served as its president in the 1890s. On my return to Quito I learned that Cordero's grandson still resided in Cuenca and would probably have given me a copy of his grandfather's book had I asked him for one. I did receive one, however, for he graciously sent me a copy later in response to a letter.

To get to Llacao we went by a local bus, a forty-minute ride over a rather bumpy road. The seats, as in all the older buses in Ecuador, are close together, so my six-foot, four-inch frame barely fit. Llacao is a small village with a large dirt central plaza with an unfinished church on one side and a small cantina on the other. Our arrival caused some curiosity, probably because they did not get to see many North Americans. My wife soon made friends with the small children, and as I watched a volleyball game on one side of the plaza, one of the players greeted me in English. It turned out that he had recently returned home after working five years in Chicago. I soon learned where the totoras grew, and after a short walk from the plaza I found a small, nearly rectangular plot of totoras surrounded by cultivated fields and pastures. The plants looked no different from the others that I had seen. They were in flower, but it was still too early for achenes. No one could tell me anything about the origin of the totoras, but I wondered if they could not be the same plot that Cordero wrote of half a century earlier. In fact, they might be the same plant that was originally introduced. The introduction of a single rhizome could spread to fill an area, constituting a clone.

A few days later we were in Lima, Peru, and my friend there, Miguel Holle, a plant breeder, proved particularly helpful. Although centuries earlier people fished off the coast near Lima in reed boats, such practice had long since disappeared. We were interested in seeing if totoras still grew nearby in swampy areas, and although we found many other sedges and cattails, there were no totoras. So my wife and I set off for the north coast, where Clinton Edwards a few years earlier had observed totora craft and the cultivation of totoras. To get there we flew to Trujillo, a lovely colonial city with many old buildings and one of the largest plazas that I had ever seen in all of South America. Early the next morning we took a bus to Huanchaco, a small coastal village, where we saw the fishermen in their totora boats, here called *caballitos* (again meaning "little horses"), much as the chroniclers had recorded four centuries earlier. As the fishermen came to shore, I was able to examine the boats and discuss their construction with the owners. They were, as I had already judged

Caballitos at Huanchaco, Peru.

from photographs that I had seen, more carefully constructed than those at Lago San Pablo. They were eleven
to twelve feet long, made from two large bundles of totora
reeds which then were lashed together, after which a small
area was cut out at the blunt end to allow the fisherman to
kneel in it.

Walking along the shore north of town, we soon came to
the area where the totoras were cultivated. About a quarter
of a mile from the sea, holes or *pozos* had been dug to a
depth of six or seven feet; there the rhizomes had been
planted. Most of the *pozos* were rather small, about fifteen
by twenty feet, although some of them were as long as forty
feet. Although the water in them was somewhat salty, the
totoras were thriving. Most of the plants were not in flower
or fruit, but I found some old flowers, and a few of them

had achenes in them, as I expected they would, for some
months earlier, D. N. Smith, a Peace Corps volunteer in
Peru, had sent me a few achenes from there.

In his visit to Huanchaco, Edwards learned that the people had formerly obtained their totora at Chan Chan, but
when the price was increased, they started their own cultivation. Chan Chan, or Tschudi, was a large pre-Inca city,
and many of the adobe buildings lasted well into this century. Chan Chan is between Huanchaco and Trujillo, and
I spent a morning there looking for totoras. Inside of what
had been the old walled city I observed rectangular ponds,
perhaps the remnants of the old water supply or sunken
gardens of the city. They were largely occupied by cattails,
and no totoras were to be found. I finally located a single
clump just south of the city in another pond dominated
by cattail. It occurred to me that it might have been the
gradual disappearance of the totoras at Chan Chan that
led to the increase in price. Also I wondered if the cattails
might not be responsible for the decline in the totoras, for
here and at other places where one might expect totoras
I found only cattails.

Sometime after I had returned to Bloomington, I found
a recent anthropological paper dealing with excavations
near the area I had visited. Old sunken gardens were
found which were from 230 to 38,250 square meters in size.
A pollen analysis revealed the presence of maize, other
crops, and totoras. Thus, it seems clear that the *pozos* were
used as gardens for growing crops. The authors believe
that the totora was cultivated, and while this is quite possible, another explanation suggests itself. Pollen of various
sedges, cattail, and pondweed was also found. These could
be weeds that found the sunken gardens an excellent habitat and came in uninvited. The same might be said of the
totora. Thus, while the archaeological discoveries prove
that totoras were present, it does not reveal whether they
were introduced and cultivated by the people. Exact dates
are not available for these sunken gardens, although the
authors consider them to be relatively late, possibly in

Totoras cultivated in *pozos* at Huanchaco, Peru.

Late Horizon times (A.D. 1476–1534) or even in early Spanish Colonial times. Thus, I still do not consider that we have proof that totoras were cultivated on the Peruvian coast in prehistoric times.

At Huanchaco I found that some of the sunken gardens had windbreaks made of totoras on the seaward side. These were not woven like the mats in Ecuador but were simply made by twining. The cut totoras had been lined up, and a string had been woven through them near the top, in the center, and at the bottom. This type of construction would be much more rapid than weaving. I was later to learn that twining was a rather common way of making mats in Peru, and I was to see many such near Lake Titicaca. I did, however, see woven mats in markets at Trujillo, and I was told they came from the sierra, most likely Cajamarca.

From Trujillo we flew back to Lima and then on to Cuzco. The former Inca capital, Cuzco, is at about 11,500 feet. We had visited it before to see the magnificent Inca stone work and to go to Machu Picchu. This time my sole purpose was to collect totoras. After arriving at our hotel, we were given a cup of coca tea "to prepare us for the altitude." That should have been a warning. After having spent nearly two weeks on the coast, we should have taken it easy our first day, but we didn't. I went immediately to the university to find a botanist who could take us to totoras. That did not take long, and I soon found myself examining totoras. As I expected, they were quite different from the ones we had seen at lower altitudes, being most readily recognized by the more condensed flower cluster and the narrower and more triangular stems. After collecting seeds and rhizomes, we returned to our hotel. It was then that we regretted our haste to get the totoras, for we realized that we had *soroche,* altitude sickness. We did little the rest of the day except to purchase our train tickets for Puno.

The trip from Cuzco to Puno takes about twelve hours, and the magnificent scenery soon made us forget our headaches. Although the train goes up over fourteen thousand feet in places, we suffered no further ill effects from the altitude. The first class car was almost filled with a tour group from Germany, and this caused me a little concern, for I had told my wife that we would have no trouble in getting a room in the pleasant, but rather gloomy, tourist hotel in Puno. After our arrival my concern proved justified, for there was no space available at the tourist hotel for our first night; the hotel clerk suggested the Hotel Lima about a block away. I have stayed at many second-class hotels in my travels in South America, and the Lima proved to one of the worst. We were awakened in the middle of the night by the police asking to see our passports. I was too groggy at the time to ask the reason why, but the next day I learned that they were looking for an escaped convict, and they searched only the cheaper hotels.

After our experience in Cuzco we decided to take it easy at Puno, which is at the altitude of 12,500 feet. It is on the shore of Lake Titicaca, so I didn't have to wait to see totoras, which I knew would be like those at Cuzco.

We first visited the market, which was not difficult, for market stalls lined the area around the tourist hotel. The stalls had roofs of totora mats, the twined variety. Also I was pleased to see totora "roots" for sale much as Cobo had observed centuries earlier. Although most writers speak of them as roots, they are actually the bases of the leaves. The leaf bases, which grow underwater, are white and somewhat succulent, and after the epidermis is peeled away from them they are eaten raw. Some have described them as rather insipid, but one anthropologist is most elegant in their praise, stating that "in tenderness and delicacy of flavor [they] far surpass the best hearts of celery." I found them rather insipid myself. Lake Titicaca was the only place where I observed the totoras used for human food. I have found no evidence that the other race of totora is ever so used. My sampling has shown that its "roots" are just as palatable.[5] It could be that only at Lake Titicaca were the people so desperate for food that they had to resort to eating totoras.

The Uros about whom Acosta wrote are extinct or nearly so, but Indians still live on islands in Lake Titicaca, and their existence there is only little better than that of the Uros of Acosta's time. The tourist brochures speak of "floating islands," but so far as I can learn, they do not float. They are, however, made largely of totora reeds, and as these decay, fresh layers of reeds are laid down. We arranged with another couple to rent a boat to visit some of the islands. Unfortunately, the motor stalled halfway there, and we sat in the boat for more than an hour before the driver could get it started again. Although the air was cool—often it is downright cold there—the reflection of the sun from the lake gave me one of the worst sunburns I have ever had.

Probably nowhere in the world is the totora more exten-

---

[5] However, maybe my conclusion is not justified. A vole once got into our greenhouse and feasted on totoras. He (or she) ate only those of the Lake Titicaca race.

Esteras used as roofs in market stalls in Puno, Peru.

sively used than by the people on those islands. Not only are their islands made of totoras, but so are their boats and their houses, the "roots" are eaten, and the leaves serve as their chief fuel—they are much used but not quite in the thousand ways Acosta indicated. The islands vary considerably in size, and the largest one we visited had a population of about one hundred and had a church—made of wood—on it and a *futbol* field. As one walks on the island the ground has a springy feel, and I would have enjoyed seeing how the football players fared on it. Near the edges one has to be careful or his feet will break through the surface and get wet.

The boats or balsas of Lake Titicaca are well known through the photographs of them that are frequently seen in articles in *National Geographic* and in tourist brochures for Peru and Bolivia. As can be seen from the photographs, they are even better constructed than those on the coast. Ordinarily they are occupied by a single person, but they

Balsa on Lake Titicaca. Stands of totora may be seen in the background.

will hold more as well as considerable cargo. There are reports of much larger craft from earlier times that would hold several people. Although I have seen photographs of balsas with sails, made either of cloth or totoras, I have never seen such. The balsas are used for fishing, hunting ducks, and making trips to other islands or to the villages for supplies. Large wooden vessels are also in use today for the latter purposes. One of these was unloading while we were on one of the islands, and a little girl fell overboard. There was considerable difficulty in rescuing her. The people there have never learned to swim, although they are completely surrounded by water, probably because the water of Lake Titicaca is far too cold for bathing.

As can be seen from the earlier quotations, the word *balsa* has been used for the boats here since early times. I thought that it was of Spanish origin, but George Carter in an article on prehistoric contacts of China with the Americas points out that balsa is very similar to the Korean word *palsa* used for water craft and implies that it comes from the Orient. Perhaps the word *balsa* deserves further study, but Professor John Dyson of the Indiana University Spanish Department verified for me that its use for boats in Spain precedes the time of Columbus. Inquiry at Puno did reveal that a Quechuan word, *wampu* or *huampu,* was sometimes used by the Indians for their balsas. I also learned that the mats, generally called *petates* in Peru, were called *q'essana* or *qquessana* by the local Indians.

I was able to get an abundance of seeds and rhizomes in the islands, so I considered the trip most successful. From Puno we took the train to Arequipa, at times traveling at altitudes over fifteen thousand feet. At that high altitude one sees no cultivated plants; in fact, one sees few plants at all, but we did see many herds of llamas, which are the livelihood of the few people who live in this rather bleak environment. I found no totoras at Arequipa, so this ends the account of my travels. The prospect of travel and fieldwork was one of the main reasons I decided to become a botanist forty years ago. Although laboratory work has occupied more and more of my time since, the fieldwork still holds its original fascination. Now to tie up some loose ends about totoras.

Two different totoras have been discussed: the totora of Ecuador, coastal Peru, and some parts of highland Peru (*Scirpus californicus* subspecies *californicus*) and the totora of Cuzco, Lake Titicaca, and a few other places in highland Peru (*S. californicus* ssp. *tatora*). Are these actually distinct species as some botanists have considered them, or are they merely races, or subspecies as they are treated here? There is no simple answer, and such situations are not as unusual as some nonbotanists might think, for if the change from race to species is a gradual process, we would expect to find some more or less intermediate situations in nature. The two totoras do differ by morphological characters that seem to be constant. The taxonomist has no rule that tells him how many such differences are needed to make a species, and sometimes closely related species may be distinguished only by very slight differences. Upon my return

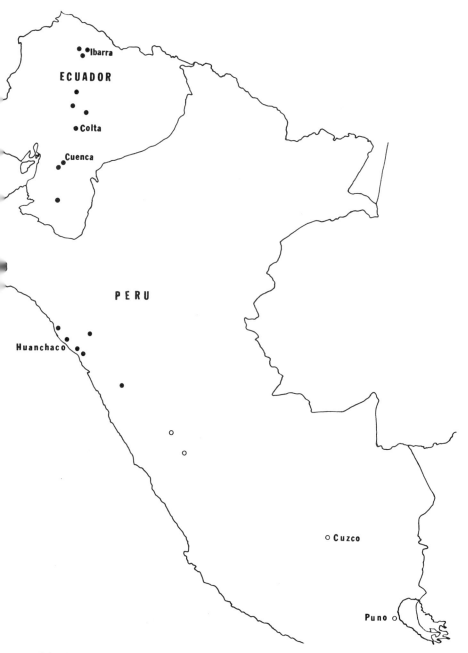

Distribution of totora in Ecuador and Peru. Dots = common totora (*Scirpus californicus* ssp. *californicus*); open circles = highland totora (*Scirpus californicus* ssp. *tatora*).

I found that the two totoras have a different number of chromosomes, and in most families a difference in chromosome number usually, but not always, indicates that one is dealing with different species. In the sedge family, however, it is not unusual to find more than one chromosome number within a single species. The two totoras have different geographical areas, but this is usually true of both races of a single species and closely related species. If the two plants were found growing together, we might have an answer to our question, for if they did so and remained distinct because of a failure to hybridize, we could be fairly sure that we were dealing with distinct species. One way to get a partial answer to our question would be to see if artificial hybrids could be made between them. My attempts at such in the greenhouse have been unsuccessful, but I don't take that as a definitive answer, for I am not sure that my technique was adequate, because the very small self-pollinating flowers make attempts at hybridization rather difficult. Also, even if two plants can be hybridized artificially, it does not necessarily mean that hybridization will occur in nature. In the final analysis all that I can say is that the decision whether to call these totoras species or races is somewhat arbitrary and a matter of personal preference.

The totora is "cultivated" in some places today in that it was introduced by human beings, but it receives little or no care, so it is not cultivated in the usual sense in which that word is used. In records from the seventeenth century in Peru there is the mention of the sowing of seeds of totora at two sites. If this is meant literally, it is somewhat surprising, for all other accounts indicate that rhizomes are used. I have found, however, that the seeds will germinate, and it seems likely that birds may be responsible for the dispersal of the plant in nature by means of achenes. The archaeological presence of the totora at a site in northern coastal Peru may stem from an intentional introduction by men, but other agents, such as birds, can hardly be ruled out. Totoras certainly grew in coastal Peru in pre-

historic times, as Heyerdahl supposed, but whether they
were intentionally cultivated is still not certain. It is clear,
however, that even if they were cultivated, the totora never
became a domesticated plant. Domesticated plants generally come to differ from their wild ancestors as a result of
artificial selection, and often they become entirely depen-
dent on people for their perpetuation. The totoras that
are found in coastal Peru and in "cultivation" in Ecuador
do not differ from the wild plants of the species.

The Lake Titicaca race of the totora (*Scirpus californicus*
ssp. *tatora*) occurs only at very high altitudes in Peru and
Bolivia, but Koyama also reports it near sea level in southern Chile. The other race (*S. californicus* ssp. *californicus*)
is found at sea level in California, at higher altitudes in
Central America and tropical South America, and again
near sea level in Chile. It is not unexpected to find these
plants at lower elevations in the temperate zones and at
higher elevations in the tropics. The latter subspecies,
however, is also found on the coast of Peru, as well as at
12,000 feet—a rather remarkable altitudinal range for any
species, whether cultivated or wild. In Ecuador at 10,500
feet and in Peru at the higher altitude the plants do not
set seed. It could be that the cooler climate at these heights
inhibits seed production. A case might be made that the
presence of the totora on both the coast and above 10,000
feet in Peru is due to introduction by human beings. I
rather like that hypothesis, but still I do not feel that we
can rule out the possibility of the introductions by birds.
If it could be shown that seeds of totoras will not germinate and survive on the coast and above 10,000 feet, the
case for the human introduction of rhizomes would be
very convincing.

One thing that can be said definitely is that the totora
of Easter Island is the same as the one in coastal Peru.
After growing and studying the plants from Easter Island
I can find no basis for continuing to recognize them as a
distinct variety. Thus one can accept Heyerdahl's claim
that the two plants are the same, but this does not mean

that the plant had to come from Peru to Easter Island. It could have arrived there from many places in the Americas, from California to Chile, or even from Hawaii. Finally, how did the totora arrive at Easter Island? I can add little to what has already been said. When I first read Heyerdahl's account, I wondered if a totora rhizome could have survived a long ocean voyage. I now feel that it could have, for I once discarded two pots of totoras in the greenhouse, and they remained under a bench for two months without any water and very little light. At the end of that time I found not only that the plants were still alive but also that they had produced new shoots. The totora is a very tough plant. Also I wondered why people would have carried totora rhizomes and not other plants, particularly seeds of food plants, on their hypothetical voyage. Heyerdahl does postulate that they carried one other plant with them, a knotweed, *Polygonum acuminatum*, which had medicinal value. He reasoned that since this plant also reproduces only vegetatively, it would have to have been introduced by man. Although this is a widely distributed species, I have not been able to verify its occurrence in Peru. Moreover, I find that it is described as producing fruits, and it thus may also reproduce by seed. I do not think its presence on Easter Island adds support to Heyerdahl's case.

Insofar as I am aware, Heyerdahl has never postulated the type of vessel that would have made the trip to Easter Island, but if it were made of totora reeds, not an impossibility, perhaps seeds of totora were carried unintentionally in the reeds. Although the seed clusters were removed before the boats that I have seen were made, it would still be possible that a seed might be present in them. Continuing with wild speculation, I might point out that perhaps an unoccupied *caballito* from coastal Peru containing a seed drifted to the island.

All of this brings me back to birds, for I still consider that a bird is most likely responsible for introducing the totora to Easter Island. Only a single seed would have been necessary, for the totora is self-pollinating, and if a single plant were established on the island it could reproduce by seed as well as vegetatively. To strengthen my case it would be desirable to name the bird that was responsible. and this I have not attempted to do. Ornithologists to whom I have talked do not think it unlikely that birds do make such trips, but they have not come up with the name of a species. So I leave it to my readers to come up with a candidate, but even if they cannot, I am reluctant to give up my hypothesis. A single bird, perhaps blown off its normal course, could have been responsible; and, of course, birds would have had thousands of years to bring about the introduction.

CHAPTER 3

Little Oranges of Quito

The *naranjilla* is a most striking plant. Its huge, dark green leaves with purple veins, purplish stems and buds, and orange fruits, all of which are rather densely pubescent or fuzzy, make it most distinctive. When the Spanish came to the New World, they called it naranjilla, "little orange," for at maturity the hairs rub off the fruits, and they resemble small oranges. In 1793 the plant was given the scientific name *Solanum quitoense* by Jean Babtiste Pierre Antoine de Monet de Lamarck, who is better known for his theory of evolution and zoological work than for his many botanical contributions. The description of it was drawn from plants grown in Paris, but Lamarck was aware that it came from Quito, which at that time referred not only to the city but also to the country that later became Ecuador. Naranjillas may have been sold in the city of Quito at that time, but the city's altitude, over nine thousand feet, is much too high for the ordinary cultivation of the plant. I did, however, observe it growing in Quito on my first visit to the city in 1962—in pots in the lobby of the Hotel Quito. I was told that it was grown there because it was considered typical of Ecuador. It is certainly one of the country's favorite fruits, and deservedly so.

This was not my first sight of it. I had previously seen it cultivated in conservatories in the United States, and I was aware of the use of the fruits. When Jorge Soria had come from Ecuador to study at Indiana University in 1956, he had proposed to do research on the species and its relatives, but he had had trouble getting the seeds to germinate and had switched to another group of *Solanum* for his doctoral study. I kept the problem in mind when I went

The naranjilla in fruit.

to Ecuador. I knew that the plant was known only under cultivation, and I was particularly interested in trying to find its progenitor. Although I wrote of the plant in some detail in my little book on nightshades some years ago, I have learned a lot more about it since that time, and I feel that it is deserving of more lengthy treatment, although I must admit at the outset that I have yet to discover its ancestor.

The genus *Solanum* is a large one, with perhaps as many as two thousand species. It is, in fact, one of the largest genera of flowering plants. Large genera are unwieldy to work with, and for that reason they are often divided into subgroups, such as subgenera or sections. When I began my work, the section to which the naranjilla belonged was ill defined, but I was able to assemble the six species that were most closely related to the naranjilla, chiefly through my own collecting in Central and South America. Now, owing to the work of William D'Arcy and Michael Whalen, we know that the naranjilla belongs to section Lasiocarpa along with twelve other species. The species of this section are largely concentrated in northwestern South America (but two occur in Asia, as we shall see) and generally are perennial shrubs, with quite large leaves, white flowers, and orange or yellow fruits, and with various parts of the plants, including the fruits, having stellate (starlike) hairs. All of the wild members of the section are also characterized by having rather prominent prickles on their stems and leaves.[1] The domesticated forms, however, lack the prickles entirely or usually have only rather small ones, probably because of selection by human beings. Anyone who has worked with the wild species will understand why, for the prickles are quite objectionable. Many is the time I have had to use my dissecting microscope to facilitate the removal of the nearly colorless spines from my fingers.

Considering its large size, the genus *Solanum* has not furnished a large number of economically important plants, but it gives us the Irish potato, which probably ranks next in importance after the three principal cereals. The section Lasiocarpa, however, provides four species that have been domesticated for their fruits, and, in addition, there are wild species that should be considered for domestication.

The naranjilla is, to my way of thinking, the best of the section—for beauty as well as taste. The plant has long been cultivated in Colombia as well as in Ecuador at altitudes of 2,300 to 6,000 feet, and in the last quarter of a century it has gained a hold in both Panama and Costa Rica. The fruit is used for its juice, drunk either directly or after adding water and sugar, or it is used to prepare

[1] I have referred to these as spines in other writings, for somehow the word prickles does not seem to do justice to them. However, by definition a prickle is a small, slender, pointed outgrowth from the epidermis, whereas a spine or thorn is a woody outgrowth derived from a branch or other plant part. Some readers may be surprised to learn that roses have prickles, not thorns.

The leaves and stems of the wild species related to the naranjilla are stoutly armed with prickles.

other beverages or various kinds of desserts. Although the rind of the fruit is orange, its flesh is green and gives a greenish juice. Some people find the color rather strange for a beverage, but most Americans like it upon their first taste. The flavor is unique, but it has been described as something like a mixture of pineapple and strawberry. In addition to its fine flavor, it is also a good source of vitamin C.

The Campbell Soup Company spent considerable effort and money in an attempt to introduce naranjilla products to the United States. Initiating the project in 1963, they made a study of cultural practices and processing in the tropics. Formulas were developed for different cocktail mixtures, blends of the juice, jams, and jellies. Commercial field plantings were undertaken in Guatemala, where the puree was canned. A blend with several other juices was test-marketed in the United States and found to be acceptable, but in 1972 the project was abandoned because the cost of the juices was too high to compete favorably with cheaper drinks having only small percentages of fruit juice. In the United States most "juice" drinks are consumed by children, who are attracted by sugar and flavor and do not care if it comes from artificial or natural sources. Perhaps with the present growing demand for natural foods in the United States, another effort will be made someday to bring naranjilla juice to this country.

The other American domesticate in the section is *Solanum sessiliflorum*, known as *cocona, cubiu,* or *tupiru.* The fruits of the cocona, which get much larger than naranjillas—to four inches in diameter—are also used for their juice, fresh in salads, or cooked in meat or fish dishes. The plant is rather widely cultivated—usually only a few plants near habitations—in the Amazon Valley from Venezuela to Peru. In contrast to the naranjilla, which usually grows above two thousand feet, the cocona grows from sea level to sixteen hundred feet.

Although the wild ancestor of the naranjilla has not been identified, there is a wild plant that qualifies as the progen-

Fruits of naranjilla being packed for trip to market near Zamora, Ecuador.

itor of the cocona. This plant, known only from southern Colombia and northern Ecuador east of the Andes, is very similar to the cocona except for being smaller in all parts and armed with prickles. Moreover, the fruits of the cocona may be orange, red, or chocolate in color and vary from obovoid to globose, whereas the wild plant always has orange, globose fruits. The cocona and the wild plant were found to produce fertile hybrids. Originally described as a species, *Solanum georgicum,* the wild plant is now considered merely a variety of *S. sessiliflorum.*

It was quite by accident that I discovered another wild species belonging to the section Lasiocarpa. Whenever I couldn't see any naranjillas in markets in Colombia, I would ask the vendor if they had *lulos,* the name by which the plant is generally known in Colombia, and in a market at Popayan I was shown some fruits that did resemble naranjillas except for being slightly smaller and lighter in color. I purchased some, and when I opened them to extract the seeds I found that the flesh was yellowish to almost white, so I was fairly certain that they were not naranjillas. Subsequently I learned that fruits of this *lulo* were collected from weeds that grew in various parts of Colombia and are used as a substitute for the real thing. Later, when I began to write up the results of my study, I could not find that this species had ever been named, although it was rather common in Colombia and was sold in many markets. So I named it *Solanum pseudolulo,* or "false lulo." It is not recommended that one combine languages, as I did with Greek and Quechuan, to coin specific epithets, but there is no rule that says one cannot. I rather like *pseudolulo*—it is most descriptive and has a certain ring to it as well.[2]

Of the strictly wild species, by far my favorite as far as fruit flavor goes is *Solanum pectinatum.* It is also a rather attractive plant if one does not object to prickles. The fruits at times are almost as large as those of the naranjilla, the hairs rub off the fruits more readily, and the taste is somewhat sweeter. The species has a rather wide distribution—southern Mexico and from Costa Rica to Peru, from sea level to five thousand feet. Some people gather the fruits from wild plants for making juice, but it may also be cultivated in some places. I feel that it deserves a wide use, and with a little effort it might become as popular as the naranjilla.

Another wild species with a pleasant flavor is *Solanum vestissimum* of Colombia and Venezuela, where it grows at higher altitudes than the other species of the section. It is a small tree and has fruits about the size and shape

---

[2] Later, in reading Linnaeus's *Critica Botanica,* I found that he had this to say: "For some past time the more critical botanists have given up names beginning with Pseudo . . . lest anyone should dub them Pseudobotanists." He was referring to generic names in this passage, but perhaps he meant it to apply to species epithets as well. However, I still like *pseudolulo.*

Fruits of naranjilla for sale at the market in Ambato, Ecuador.

of a duck's egg. The chief objection to eating it is that the
hairs on the fruit are quite bristly and difficult to remove.
Two different people whom I encountered while collecting
in Colombia told me that the fruit was sometimes eaten,
and they volunteered the information that it also was used
in the treatment of high blood pressure. I haven't been
able to learn more about the latter use, but I hardly suppose that it can be a very ancient one, for I doubt if the
Indians had heard of high blood pressure until fairly
recently. The material that I had seen of this species in
herbaria was labeled *S. quitoense,* but I knew that it was
different from that species. After a search failed to turn up
a name for it, I described it as a new species, *S. tumo,*
using one of its native names for the specific epithet. My
search, however, was not as thorough as it should have

been, and only after the new name was published did I discover that it had been described over one hundred years earlier as *S. vestissimum*. It is embarrassing to admit the oversight, but such has happened to many other people who have given names to plants.

After I had assembled all of the above-named species and two others, yet to be mentioned, I began my study of the group. I proposed to make a detailed morphological study to see which species most closely resembled the naranjilla and to attempt reciprocal crosses among all the species for what information that might give me about relationships.

The first part of the study gave very clear results. There was no question but that *Solanum candidum* was the species most similar to the naranjilla. I reached this conclusion simply by inspection of the various species, and it was borne out later by various kinds of numerical taxonomic studies. *Solanum candidum* is widely distributed, ranging from Mexico to Peru, and most commonly known as *huevo de gato* ("testicle of the cat"). So far as I can determine, the fruits are only rarely eaten by human beings. The flavor isn't particularly good, and the hairs are very difficult to remove from the fruit. There are other differences between it and the naranjilla, but if people had brought *S. candidum* into cultivation—why, I find rather difficult to imagine—they might have selected for deciduous hairs and better flavor. The naranjilla also has larger leaves and fruits and has small prickles or none at all. These differences might readily be accounted for by human selection. However, the fruit flesh is yellow or orange in *S. candidum* and a dark green in the naranjilla, which would be difficult to explain if *S. candidum* were the progenitor. Although people might have been able to establish it in cultivation above its normal range, another reason that I regard *S. candidum* as an unlikely direct ancestor is that it occurs at elevations somewhat lower than those at which the naranjilla is grown.

The results of the crosses do not entirely eliminate

S. candidum from consideration but provide no strong support for it, either. Crosses between most of the species failed. In fact, only four were successful. Three of these involved the naranjilla and were successful only when it was the male parent. One of these was with *Solanum vestissimum*. Fruits were produced in two of ten attempts, but their seeds were nearly flat and failed to germinate by ordinary means. After examining the seeds under magnification, I found that apparently normal embryos were present in some of them. Six were dissected out of the seed and planted on nutrient agar, and two germinated. After the seedlings were an inch tall, they were transplanted to ordinary soil and grew into vigorous plants. These F_1 hybrids produced 28 percent seeds on self-pollination, and they were difficult to backcross to the parents.

Very similar results occurred in the cross of *Solanum candidum* with the naranjilla. However, one normal seed did develop. The plant secured from it and five hybrids from embryo culture were also vigorous and had a seed set of 41 percent. Again, backcrosses to the parents were difficult to obtain.

The best luck was with *Solanum hirtum*, another widely distributed species. This cross was made readily, and 50 percent of the seeds were normal. These gave vigorous F_1 hybrids which had about 70 percent seed set, and backcrosses were readily made to the parents. Thus, on the basis of the ease of crossing and fertility of the hybrid, *S. hirtum* appears to be the species most closely related to the naranjilla. Its morphology, however, does not make it appear to be a likely progenitor, for it differs from the naranjilla in a number of characters. It also grows at a lower elevation than does the naranjilla. Moreover, I have no indication that its fruit is ever eaten. In fact, its rather small fruits have an unpleasant odor. One of my assistants strongly objected to harvesting its seeds for that reason, although I don't think the odor is all that bad. Yet, as we shall see, *S. hirtum* may have something to offer to the naranjilla in spite of its unattractive fruits.

Roots of naranjilla infested with nematodes.

The naranjilla suffers from a large number of fungal diseases and insect pests. One of the most serious is root-knot nematode, which drastically cuts yields as well as making the plant more susceptible to other pests. Because of the widespread presence of nematodes in the agricultural zones, the farmers in Ecuador try to find virgin areas for starting new plantings. Trees are cleared from forested areas, sometimes on slopes so steep that it is difficult to stand. Naturally, such areas are subject to severe erosion, but even though the local ecologists cry out against the practice, the farmers refuse to heed their warnings because the naranjilla is one of their most important cash crops. After a short time the nematodes find their way into the new plantings, through mud on the farmers' shoes or on his agricultural implements, so although the naranjilla is

Area cleared on mountainside for growing naranjilla near
Zamora, Ecuador.

a perennial and should continue to yield indefinitely, the
farmer abandons his plot after a few years and looks for
other new areas.

In 1981 very few naranjillas were reaching the markets
in Ecuador, and those that were available were naturally
quite expensive. The few that were being produced had to
travel great distances to reach the markets, first by mule
from the mountains to the roadsides and then by truck to

the markets in the cities. For the first time agronomists became quite concerned, for not only were the people being deprived of one of their favorite fruits, but the farmers who depended on it as a cash crop were suffering as well. Obviously something had to be done, and this is where *Solanum hirtum* enters the picture.

It so happened that one day in the sixties when I was discarding some of my plants, I discovered some knoblike growths on the roots of my naranjillas. I immediately suspected nematodes, and that was cause for alarm, for the nematodes are a difficult pest to eliminate from the greenhouse. My concern grew as I went through the pots of the other species and found that every one of the plants had the same irregularly swollen roots except for *Solanum hirtum.* Inasmuch as the unaffected plants were growing on the same bench as the others, I saw no way they could have escaped the nematodes unless they were resistant or immune. Although I wasn't pleased with the presence of nematodes in the greenhouse, they did provide information that I wouldn't have come by otherwise. Since *S. hirtum* would cross readily with the naranjilla, I realized that we might have something of potential significance.

Why *Solanum hirtum* should be resistant to the nematodes, and the other species not resistant, is unknown. Possibly it could be because of the presence of an alkaloid not found in the other species. If so, and if the alkaloids occurred in high concentrations in the fruits as well as the roots, *S. hirtum* would have little to offer to us, though I know of other *Solanum*s, including some in this section, that have alkaloids in the young fruit but become safe to eat at maturity. Many parts of the Irish potato, including the tubers, contain the alkaloid solanine. In fact, in the past some new varieties of the potato were released to growers before they were found to have such high concentrations of solanine that they later had to be recalled.

That *Solanum hirtum* was resistant to the nematodes in our greenhouse did not, of course, mean that it was resistant to tropical nematodes. Therefore I sent seeds of both species and of the hybrid to Jorge Soria in Costa Rica. He found that the same results held there, and, moreover, through backcrossing he was able to produce plants that were resistant to nematodes and whose fruits had a good size and flavor. Because of the pressure of other duties he was not able to continue the work.

Nematode resistance is also found in some other species of *Solanum* that will not cross with the naranjilla. The naranjilla can be grafted to rootstocks of some of those species to secure resistance to nematodes. Although chemicals now are known that can be used to treat the soil to kill nematodes, the ideal solution to the nematode problem would be to have seeds of a resistant strain of the naranjilla for the farmer. In the long run that would be less expensive and less time-consuming than the other methods of nematode control. It will take several years of concentrated effort to do the breeding necessary to produce such strains. It was impossible for me to do the kind of work required in greenhouses at Indiana University, so I was delighted to have a letter in 1981 from Dr. Saul Camacho, a horticulturist in charge of the fruit program of the Instituto Nacional de Investigaciones Agropecuarias of Ecuador, requesting seeds of my species and informing me that they were initiating a plant-improvement program for the naranjilla. Eventually naranjillas may be developed with genes for nematode resistance from *Solanum hirtum*.[3]

Other possibilities for the improvement of the naranjilla exist through interspecific hybridization. Possibly one of the species with which the naranjilla will hybridize may

---

[3] In 1983 Dr. Camacho left Ecuador to accept a position in Indonesia. His departure and financial problems at INIAP led to an abandonment of their program to improve the naranjilla. Since then, however, I have been able to set up cooperative arrangements with agronomists at Turrialba, Costa Rica, and Medellin, Colombia, in an attempt to develop nematode-resistant naranjillas from hybrids of *Solanum hirtum* and *Solanum quitoense*. From a recent letter I have also learned that INIAP is re-establishing their program.

offer resistance to some of the other pests and diseases
that plague the plant. Also, if these other species grow in
areas to which the naranjilla is not adapted, it may be
possible to develop naranjillas that can grow both at higher
and lower elevations than they do now.

Leaving practical considerations and returning to the-
oretical ones, I should point out that the origin of the
naranjilla remains to be discovered, if *Solanum candidum*
is not the progenitor. The naranjilla has never been found
growing truly wild, although it occasionally escapes from
cultivation. Perhaps truly wild ones will eventually turn
up which would prove to be the progenitors of the domesticated plants. The most likely place for them, I feel, would
be Colombia, and I would expect the plants to have rather
stout spines like the other wild species of the section, and
they would probably have fruits with green flesh. I would
also expect that they would cross readily with the naranjilla and give fertile hybrids. Such a plant could well have
disappeared, however, adding to the long list of species
that have been brought to extinction by human beings. If
so, it is a pity, for it might have preserved genes that
would be valuable in helping to control the diseases and
pests of the naranjilla.

The next part of this *Solanum* story takes us to Asia.
I was not aware that species related to the naranjilla occurred in the Old World until I went to Thailand in 1980.
In Bangkok I saw hairy orange fruits, obviously *Solanum*,
in the markets. Some bins had small ones, a little over
¾ of an inch in diameter; other bins had larger ones, over
1¼ inches in diameter. My first thought was that the larger
ones might be naranjillas, but that seemed most unlikely,
for only a few months earlier I had sent naranjilla seeds
to a Peace Corps volunteer at his request because he
thought it a plant worth trying in Thailand. After pur-
chasing some fruits, I found that the hairs on the fruit
were longer than those on the naranjilla and did not rub
off as readily. When I cut the fruits, I found the flesh
to be yellowish in the small ones, but to my surprise the

larger fruits were light green near the center. The next day I described the fruits to Somsak Sansukh, a former student of mine, and he told me they were from *Solanum ferox,* called *ma uek* in Thai, and that they were used in making curries. Later I learned from Mike Whalen that the plant generally called *S. ferox* in Asia should correctly be known as *S. lasiocarpum.*

When I returned to Bloomington, I planted seeds of the two *ma ueks* in the greenhouse. Seeds from the smaller fruits produced plants about a yard tall, and the leaves and stems were prickly, agreeing well with the description of *Solanum lasiocarpum.* Seeds from the larger fruits gave plants about twice as tall, with somewhat larger leaves and flowers, and they were completely unarmed at maturity. The larger size and the lack of spines suggested that the latter might be a domesticated form of the species, but it was to be some time before I was to have a definite answer.

A search of the literature revealed that *"Solanum ferox"* was always described as a prickly plant, and the fruits were used in India, Thailand, and Malaysia in preparing sour sauces for use in curries and also had use in medicine, but nowhere did I find any indication that unarmed forms of the species were known or that the species was cultivated. I sent off a number of letters to botanists in Thailand trying to learn more, and after several months I had an answer. Umpai Yongboonkird, another former student of mine, wrote that the small-fruited, spiny plants grow wild or as weeds, but that the larger-fruited forms without prickles occurred as "only one or two plants in the orchards." On this basis I considered that it was cultivated and must be a truly domesticated plant. It seems rather surprising that one could go into the capital of any country today and find a previously unrecorded domesticated plant, but apparently I had. In due time I described it as *Solanum lasiocarpum* var. *domesticum.*

Solanum lasiocarpum is also of interest for other reasons. It is very similar morphologically to the American *S. candidum,* so much so that there may not be justification for

regarding the two as separate species. The very close relationship is also supported by hybridization studies, for I
found that the two would cross readily and produce hybrids nearly as fertile as the parents. The progenitor of
S. lasiocarpum almost certainly had to come from the Americas, and most likely it was derived from S. candidum—but
when and how? It can hardly be a very recent introduction,
for it now has a wide distribution, from India east through
Indochina, southern China, Malaysia, and Indonesia to
the Philippines and New Guinea. Although birds are
thought to be involved in its dispersal, it seems rather
unlikely that they could have carried seeds for the vast
distance that would have been required to introduce it
from the Americas. So we have to look to human beings,
and I think that they could have acted as the carriers
as late as voyages of the Spaniards and Portuguese in the
sixteenth and seventeenth-centuries. It seems unlikely that
the fruits or seeds were carried intentionally, since S. candidum is seldom used by mankind and has little to recom-
mend it. Fruits or seeds must somehow have gotten on
board a ship accidentally. It may seem very unlikely that
such ever happened, but many weeds appear to have gone
from one continent to another as uninvited passengers on
transoceanic voyages. For S. lasiocarpum to have attained
its present wide distribution in Asia in a few hundred
years could be accounted for by dispersal in part by birds
and in part by human beings who had found a use for the
plant. Because the differences between S. lasiocarpum and
S. candidum are very slight, I see no need to call upon a
very long period of time for the changes to occur.

Solanum lasiocarpum is not the only species in the section
whose distribution presents a problem. A second species,
S. repandum, is found only on islands in the Pacific—the
Society and Marquesas islands west to Fiji, Samoa, and
Tonga—where it has generally been found as a weed near
human habitations or as a cultivated plant. The first speci-
men of it ever collected comes from the voyage of the
Endeavour of 1768–71 to the Society Islands. Thus it is

clear that the plant has been in the Pacific area for some time. *S. repandum* is very similar to the cocona and has been used in similar ways as a food on Pacific islands. Once again, the differences between the two are so slight that there may be no justification for considering them as separate species.

Since the cocona is used for food, an intentional introduction does not appear as unlikely as it did for the previous species. Such an introduction could have occurred on a prehistoric voyage from the Americas into the Pacific, an explanation that I am sure will appeal to the diffusionists (see chapter 2). On the other hand I don't think an early voyage by the Spanish after the discovery of the Americas can be ruled out. One possibility, for example, would be the Mendoza-Quiros expedition, which sailed from Piata, Peru, and landed on the Marquesas Islands in 1595. The Spanish are known to have introduced South American plants to some of the islands in their early voyages, although I know of no specific record of the cocona. The objection might be raised that if the Spanish introduced the cocona so late, there would not be time for it to differentiate into another species, but I would point out that *Solanum repandum* differs only very slightly from the cocona, and again I see no need for any great length of time being necessary. The greatest difficulty perhaps concerning a transport by humans in either prehistoric or historic times is to explain how the fruits or seeds of cocona reached the South American coast. Today the cocona occurs only on the eastern side of the Andes, and if the distribution was the same in the past as it is now, then we have to assume that people carried seeds over the Andes to the coast—not an impossibility but not very likely.

An introduction of the progenitor of *Solanum repandum* by natural means hardly offers a more acceptable explanation. The fruits of the plant certainly could not have floated very far, and it is unlikely that the seeds would have fared better. Transfer by birds seems the only plausible natural means. However, it is mostly shore birds that are known to make long flights into the Pacific, so we would still have the difficulty of explaining how they picked up seeds on the Upper Amazon. It should be noted, however, that birds were probably making long flights into the Pacific thousands of years before humans ever made the journey, so they would have had a far longer time to transfer the cocona or the common progenitor of the two species, onto Pacific islands. A transfer by birds, of course, would not explain why the fruits are used in similar ways in South America and on the Pacific islands.

Unfortunately, we may never be able to learn much more about *Solanum repandum,* for it is near extinction, if not already extinct. Whalen failed to find it in visits to some of the places where it had previously been reported, and my letters to botanists within its range did not reveal any sources. So I have never seen the plant in the living state, and it now appears unlikely that I, or anyone else, ever will.[4]

Although it does not belong to the same section of the genus as the species previously discussed and was never the object of particular study by me, some comments about *Solanum marginatum* will provide a fitting close to this chapter. This species, a shrubby plant with prominent prickles on the stem and leaves, occurs as a weed in waste places around Quito and in many other parts of the Andes. It is sometimes called *sacha* (from the Quechuan for

---

[4]Since I wrote this chapter, I have grown seeds of a *Solanum* from Fiji which I had received from Professor Richard Hamilton sometime earlier. The plants appeared rather similar to some of the others discussed in this chapter, but they lacked completely the stellate hairs found in all of the other species. It was not until Mike Whalen reminded me that some plants of *Solanum repandum* in Fiji were known to lack these hairs entirely that I realized that we had this species. Thus, although the species may be extinct in the Pacific islands, we now have it in "captivity," and it will be possible to learn more about its relation to the other species. At this writing I have fruits developing in the greenhouses from crosses of this species to a number of other ones. To me one of the most impressive things about the plants is the strong aroma of the ripe fruit; in fact, it might be described as a sickening sweet odor.

Branch of *Solanum marginatum* in fruit.

"false") *naranjilla* in Ecuador, for the fruit is orange and about the size of those of the naranjilla although lacking the conspicuous hairs of that species. It also has another Quechuan name, *huapag,* which led me to believe at first that it was an indigenous species. I soon learned, however, that it was native to Africa and had been introduced to the Andes. Whether the introduction was unintentional or whether the species was introduced as a possible ornamental is not entirely clear. In either event it found the Andes a fine secondary home and soon became well established as a weed. The fruits are toxic, but the people learned that they could use them to produce a suds for washing clothes. The plants were also sometimes painted and used for Christmas trees in the homes of the poorer families. These were about the only uses for the plant until fairly recently. An Ecuadoran chemist (who also taught me a lot about the botany of Ecuador), Alfredo Paredes, in analyzing the fruits discovered that they contained high amounts of steroids, and he thought that the plant might prove to be a commercial source of the drugs. He was correct, and today the Schering Company of West Germany is using it to produce solasodine.

The Ecuadoran operations are under the direction of Leo Roth. Near Quito he has experimental plantings, including collections of *Solanum marginatum* from many parts of the world as well as many other species of *Solanum.* A special plot is infested with root-knot nematodes, for it is also a serious pest of the plant, and the different species are tested for resistance to nematodes. The commercial plantings are at Cotocachi, near Otavalo, where Otavalo Indians are hired to harvest the fruits. The fruits are then taken to their plant at Quito, where the solasodine is extracted for shipment to Germany. There a small amount is used for the manufacture of birth-control pills and a larger amount goes into the production of anti-inflamatory drugs.

On my visit to Ecuador in 1982 I was able to visit Cotocachi. I learned that although the plant does well as a weed, particularly in those places used by the people for their toilet, its cultivation was not a simple matter, and after eight years they are still improving the ways of growing the plants. In cultivation the plants are still no different from the weeds in Quito, but changes can be expected. Plants with the greatest concentration of solasodine and the greatest production of fruits are being sought, but other characters are not overlooked. For example, plants without prickles would be a great boon. Presently the Indians who harvest the fruits are provided with stout leather gloves for protection. While I was there I was shown one plant that bore no prickles. Whether the lack of prickles on that plant will prove hereditary and possible to incorporate in other plants remains to be seen. What I would like to point out is that in this case we are witnessing a wild plant becoming a domesticated one. Since the domestication of most plants occurred in the prehistoric period, we have little record of what occurs during the process of domestication. What happens to *Solanum marginatum*, and other new plants being domesticated, should be of great interest to students of domestication.

CHAPTER 4

Chenopods: From Weeds to the Halls of Montezuma

Goosefoot, lamb's-quarters, and pigweed are all the same plant and hardly an attractive one. One writer describes the plant as "homely." It is one of the weeds with which I became acquainted as a boy, for in the spring my grandmother would gather it for "spring greens." This is one of the reasons that the plant is fairly well known, for many people still gather it for potherbs. Usually found in some abundance in most parts of our country, it is easily collected, and when boiled and seasoned, it makes a fine dish for those who like such vegetables. A great deal about its use as greens, and also about its fruits, may be found in Fernald, Kinsey, and Rollins's book on edible wild plants, which, though it is some years old, I still consider to be the best book on the subject. The genus name of the plant is *Chenopodium,* from the Greek for "goosefoot," in reference to the shape of the leaf of some of the species, and it brings its name to the family, the Chenopodiaceae. It should not be terribly surprising that it makes good cooked greens, for to the same family belong spinach and Swiss chard, both long ago domesticated in the Old World as food plants.

There are about fifty species of chenopods in the United States, of which ten are unintentionally introduced species, coming from such diverse places as Europe, Asia, Latin America, and Australia. One of the introduced species bears one of my favorite plant names, Good King Henry (*Chenopodium Bonus-Henricus*). Although it is almost impossible to trace the origin of many common names, the herbalists give a fairly good account of this one. Origi-

Lamb's-quarters (*Chenopodium missouriense*).

nally it was thought to be a kind of mercury, a plant now recognized as belonging to a completely different family, and it was called Good Henry because of "a certain good quality it hath," as the herbalist Gerald tells us, to distinguish it from Bad Henry, "a certain pernicious herb." The "King" part apparently was a later addition, but *Henricus* did not originally come from King Henry, for Maud Grieve in her *Modern Herbal* (1931) tells us that Henricus in Germany refers to elves and kobolds. The plant, like most, has other common names, one of which is fat hen, because chickens would readily eat it. In England it was formerly widely cultivated as a potherb.

Chenopodium album is another of our introductions from Europe, and until recently that name was also used for some of our native species. The classification of chenopods is no simple matter, and only in the last few years have the American species become well recognized, one reason being that some of the best characters reside in the fruit and seed,[1] which were largely neglected by earlier taxonomists. Another introduced species, also now a weed in many places, is Mexican tea or wormseed, *C. ambrosioides,* which came to us from Latin America. It is strongly aromatic, and one is unlikely to confuse it with the other species. The odor is difficult to describe; some people find it pleasant, others do not. This species is noteworthy for its fruits, which have long served as an efficient vermifuge. A tea made from the plant is also used for other ills in Mexico, and it is widely used as a spice in a number of dishes there.

Although the names goosefoot and lamb's-quarters are likely to cause little confusion, the same cannot be said for pigweed, for this name is also commonly used for some members of the genus *Amaranthus.* Chenopods and amaranths are rather similar in many ways. Both have small flowers which lack petals, and the flowers are rather densely aggregated into flower clusters or, more properly speaking, inflorescences. The best way to distinguish the two is to examine the flowers and inflorescences. The amaranths have bracts subtending the flowers, and the calyx will be dry and membranaceous, whereas the chenopods lack the bracts, and the calyx is green and often somewhat fleshy. The small size of these structures may make it difficult for nonbotanists, and some botanists as well, to tell the plants apart. Fortunately, one can sometimes learn to distinguish them readily without reference to the flowers. Many of the chenopods, particularly the edible ones, have a white mealiness on their leaves, particularly the younger ones, which is not present in the amaranths. The leaves of the chenopods are also usually more fleshy than those of amaranths, and the veins are not as prominent. The ama-

---

[1]The fruit is a rather small, one-seeded structure, generally called an achene. At maturity the fruit is mostly seed, the ovary wall remaining only as a thin layer. Nonbotanists would likely refer to the fruits as seeds.

Quinua and Sergio Soria.

ranths are currently enjoying considerable attention in the United States as potentially valuable food plants, but neither they nor the chenopods are new food plants. Three species of both genera were brought under domestication by the Indians of the Americas in prehistoric times—not primarily as potherbs but for their fruits, which were used as a cereal. Since the name *cereal* is reserved for members of the grass family, the chenopods and amaranths, along with buckwheat, a plant of Old World origin, are often referred to as the pseudocereals.

My first trip to South America in 1962 was primarily to study the *Capsicum* peppers, but a secondary purpose was to become acquainted with other Andean domesticates, many of which, including the chenopod known as *quinua* (*Chenopodium quinoa*), I had never seen before. My first

contact with it was at Otavalo, Ecuador, where the fields
of it presented an array of colors—pink, red, violet, and
yellow, as well as green. I thought them a beautiful sight,
and I didn't find the plants at all homely, though some
fields had all green plants, and I must admit those were
not terribly attractive. I was traveling with my friends,
Jorge Soria and Jaime Díaz, and they asked the owners
of a field, a man and his wife, to allow us to go into it
for a closer examination. The owners not only consented
but also gave us a personal tour, and the man pointed to
some plants and told me that they were *ashpa quinua,* which
for our purposes we can translate as wild or weed quinua.
These, I learned, had black fruits in contrast to the white
fruits (they called them *semillas,* "seeds," of course) of the
domesticate, and the man tried to eliminate the young
plants from the garden. How one can tell the two apart
before the fruits are mature is no simple matter. I later
learned that *ashpa quinua* was frequently a companion
weed with the crop, and I was also to find it in cornfields
and other places. I realized at the time that there were
interesting and important problems for study in the chenopods. Since, unfortunately, it has not been possible for
me to give detailed study to quinua and many of the other
Andean plants that I found interesting, I did the next
best thing: I turned the problems over to students. On this
trip and on later trips I collected fruits of chenopods, both
wild and domesticated. These were later to contribute to
the doctoral research of two of my students, David Nelson
and Hugh Wilson. The former became allergic to the pollen of chenopods, so he did not continue his study of them
beyond the doctoral degree, but the latter is still contrib-
uting much to our knowledge of the origin and relationships of the domesticated and wild chenopods.

In addition to quinua, another species, *cañihua* (or *kañiwa*), *Chenopodium pallidicaule,* is also grown to a limited
extent in the very high Andes. In Mexico still another
species, *huauzontle* (*C. berlandieri*) has become domesticated,
and in prehistoric times *C. bushianum* was an important

food source in the eastern United States. Whether the last was wild or cultivated is not entirely clear. Let us begin the story in South America.

The archaeological record tells us that people have lived above ten thousand feet in the Andes for a long time, and we know that some civilizations developed at those high altitudes long before the Incas. Life, to us at least, appears rather harsh there for both people and plants, and only a very few domesticated plants do well. Maize can be grown at ten thousand feet, but it is not very productive at that altitude, and the chief crop became the Irish potato.[2] Other tuber crops were also domesticated, so there were good sources of carbohydrates, but they furnished little in the way of protein. The chenopods have largely filled that role. Some people also had animals, alpacas and llamas, but the animals were so valuable in other ways that many people still seldom use them for food. With the coming of Europeans, some of the cereals, chiefly barley, were found to do well in the high Andes, but the chenopods will grow still higher.

Presently quinua is largely confined to Ecuador, Peru, and Bolivia at altitudes from about 6,500 to 13,000 feet. Some is still grown in southernmost Colombia, in the Andean part of Argentina, and at lower altitudes in Chile. Attempts to introduce it to other highland parts of the world have met with little success. In most places quinua is grown much in the same way that it was in the time of the Incas, with seed broadcast and harvesting and winnowing done by hand. Some mechanization of the operations is now used in Peru, however.

A great number of varieties, or cultivars, of quinua are recognized. Sometimes they are grouped as bitter or sweet. In reality, all are bitter—some more so than others—because the fruits contain saponins, which are responsible

---

[2] It would be better to call it the Andean potato but the name Irish for the white potato is so firmly established that it is probably too late to try to bring about the change.

for the bitterness and in high concentrations may be toxic as well. Therefore the fruits are thoroughly washed in water before they are eaten, and the saponins come out in the form of suds. I have been told by farmers that the "sweet" varieties containing very little saponin are more subject to bird damage than are the bitter varieties, but I have seen no scientific verification of this.

Quinua is used in a variety of ways, most commonly in soups or stews or as a porridge. The curved embryos of the seeds come out after cooking and look like little worms. Referring to my notes, I find that I describe the taste as "earthy," and Fernald, in referring to products made from fruits of the wild species in the United States, describes their taste as "mousey." By itself quinua is used to make a crude bread, and mixed with wheat flour it makes a good bread. As a food it is nutritious, for the fruits contain around 15 percent protein, more than most cereals, in addition to considerable amounts of carbohydrates. Moreover, the amino acid lysine, which is low in cereals, is well represented in quinua. The grain at times has been used to make a *chicha*, or beer, and today is combined with other ingredients to make such a beverage. The young leaves are used as potherbs, and the ashes from the old stems are used to make a paste that is chewed with coca leaves in Bolivia and Peru. Apparently the calcium in the ash causes a more ready release of the active ingredients in the coca leaves.[3]

There is little in the way of archaeological material concerning quinua, so we do not know how old it is as a domesticated plant. However, it differs in several ways from truly wild chenopods. The greatest changes have been in the fruits and seeds, as would be expected, for these are the chief characters for which the plant is cultivated. The fruits of domesticated quinua are large and

---

[3]Although cocaine is made from coca, chewing the leaves does not have the same effect as taking pure cocaine. It acts as a mild narcotic enabling people to work long hours without feeling fatigue or hunger.

Harvesting quinua near Lake Titicaca, Bolivia.

light in color—white, pink, or yellow—in contrast to the
small, dark fruits of wild chenopods. The shape of the
fruit is slightly different, and the thick seed coat has been
reduced. These changes may be responsible for the rapid
and even germination of the seeds of the domesticate in
contrast to the slower and uneven germination of wild
chenopods. It has also been suggested that the domesticates
contain less saponin than do the wild types. For some that
is probably true, but since the wild ancestor of quinua has
not been definitely identified, we can only speculate for
the species as a whole. Another change that has occurred
is that the inflorescence has become highly compact in
quinua in contrast to the rather diffusely branched inflorescences of wild chenopods. The fruits ripen more or less
simultaneously in quinua, whereas they mature over a pe-

riod of weeks in the wild plants. Another important difference is that the quinua fruits tend to remain on the
plant after they ripen, while in the wild they tend to "shatter" or fall as they ripen. Finally, quinua is in general
larger in all parts, sometimes reaching heights of six feet,
than the wild chenopods.

The origin of quinua remains unsolved. It is a tetraploid,
but it is thought to have arisen directly from another tetraploid rather than by hybridization of diploid species followed by chromosome doubling. There are no wild diploids in the Andes that could have been its ancestors, but
there is one tetraploid, *Chenopodium hircinum*, that appears
to be closely related and is presently being investigated
by Wilson. It, however, is found in southeastern South
America and at lower elevations than is quinua. The other
possibility is that a tetraploid species, closely related to
the domesticated chenopod of Mexico, to be discussed
shortly, was carried to the Andes. Its introduction there
could have been by natural means—birds, for example—
or it could have come with humans, either unintentionally
as a camp-following weed or intentionally as a useful,
perhaps even cultivated plant. It seems likely that it was a
dark-fruited form, for Nelson has shown that the white
fruits of the Mexican chenopod and those of quinua are
controlled by separate genes. Once in the Andes, then,
this Mexican introduction could have given rise to both
quinua and *ashpa quinua*, the weedy quinua. The latter is
probably closer to the progenitor in that it has black fruits,
but it has changed considerably from the original ancestor
through continued crossing with the domesticate.

Although N. W. Simmonds in 1974 saw no future for
quinua other than a continual decline, I feel that the crop
may well hold its own in the Andes for a long time to
come. When I first visited the Andes in 1962, some of the
local plant breeders regarded quinua with some scorn,
referring to it as food for Indians. Their time, they thought,
would be better devoted to wheat and other European
crops, although they did realize that work with corn (an-

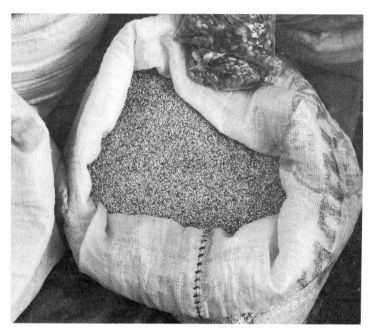

Quinua "seeds" for sale at market in Quito.

other Indian crop) was important. The attitude seems to have changed since that time, and many people realize that quinua furnishes a much-needed source of protein in the highlands. More attention is now being given to its improvement in several of the Andean countries. Quinua, originally not eaten much outside the Andes, is now also in some demand in Lima, where the highlanders, who have migrated to the city in great numbers, like their traditional foods. Experiments have shown that quinua can be grown successfully at fairly low altitudes near Lima, but it seems likely that most, if not all of it, will continue to be produced in the highlands.

Chenopodium pallidicaule, known as *cañihua* or *kañiwa*, used much like quinua, also is cultivated in the Andes but is largely limited to the altiplano of Peru and Bolivia. It withstands frost and tolerates drier soil better than does quinua and has even a higher protein content. It is a much smaller plant, however, and its yields are lower. Moreover, it is probably best considered a semidomesticate. Its fruits are black and smaller than those of quinua and more likely to shatter at maturity. It is a diploid species and not particularly closely related to other edible chenopods. It has no archaeological record, so its time of origin is unknown. Daniel Gade has suggested that it may originally have been a weed in quinua fields and that it became intentionally cultivated when it was found to do better than quinua in the very high regions. Wild forms of the species are said to occur near Lake Titicaca.

Chenopods seem not to have been used in most of Central America, and it is not until we get to Mexico that we again find they are important. Although the Mexican chenopod is a pre-Columbian domesticate, it was not given a scientific name, *Chenopodium nuttalliae*, until 1918. Many botanists have continued to think of it as nothing more than a form of *C. quinoa*. This is understandable, for the Mexican plant is very similar to quinua. Much of the similarity, however, is because in developing it people apparently selected for many of the same traits that distinguish quinua from the wild plants. It has been shown, however, that hybrids between it and quinua have highly reduced fertility, and Nelson and Wilson found that the two can be distinguished by characteristics of the leaves. Wilson has also shown that the Mexican domesticate is rather similar to the wild *C. berlandieri*, a widespread, highly variable wild species of the United States and Mexico. The two give fertile hybrids, and it appears likely that this wild species gave rise to the domesticate. It is clear that the two belong to the same species, so the name of the domesticate has become *C. berlandieri* var. *nuttalliae*.

Three subvarieties, or cultivars, of this variety are grown in Mexico today. By far the most important is *huauzontle*, of which the inflorescences containing the young fruits are eaten. These are dipped in batter before cooking and then are eaten by pulling a section of the inflorescence through

Quinua *(left)* and *cañihua (right)* in an experiment station in highland Bolivia. The man in the middle is Jorge Soria.

the teeth so that the young fruits are pulled off and the tough "strings" or stalks are left. It is a popular vegetable in Mexico City today and is sometimes seen on the menus in the better restaurants. It does not have the "earthy" flavor of quinua. An improved selection, Santa Elena, was introduced by the agricultural experiment station at Chapingo a few years ago and is now grown in some abundance near Mexico City to supply the demand.

The use of a chenopod for grain in Mexico is rare today, but the practice still exists. One might suppose it was used for that purpose before the inflorescences became popular. In late prehistoric times and early historical times the people of Mexico were required to pay a tribute to the emperor, Montezuma, in the form of agricultural products. Among these was a grain, known as *huauthi,* for which an annual payment of 160,000 bushels is recorded. There is no agreement about what the grain was. Some authorities think it was an amaranth, a few think it may have been a chenopod, and others a mixture of both cereals. The crop had a significant role in the religion of the Indians, so the Spanish tried to suppress it. If indeed it was a chenopod, then at one time it was an important grain crop, and its rarity today could be explained. The name used for the grain chenopod today is *chia,* or *chia roja* ("red chia") because of its pink or reddish seeds. This name is sometimes the source of confusion, for certain mints whose fruits are used to make a refreshing beverage are also known as *chia.* In my search for the chenopod *chia* in Mexico, I usually ended up by being offered the mint.

Apparently the only use for the grain *chia* presently is to incorporate it into a corn dough for making tortillas. I have tried to learn if the grain is washed like quinua before it is used, and Cristina Mapes informs me that, at least in the Patzcuaro area, no special preparation is given to it. I had found nothing in the literature regarding saponin in the grains, and I find that they contain very little, far less than does quinua.[4] At present one cannot predict much of a future for *chia,* but I shall have more to say about it at the end of this chapter.

The third way in which the plant is used in Mexico is for greens. Probably the other two varieties are also used

---

[4]A simple method can be used to test for the presence of saponin. By placing the fruits in water and shaking them vigorously, suds formed by the saponins will appear on top of the water. I took equal amounts of the chenopod fruits I had at the time—three samples of quinua and one each of *chia, Chenopodium bushianum,* and *C. missouriense*—and added the same amount of distilled water to each. The last showed the largest amount of suds, the samples of quinua varied somewhat, *C. bushianum* had the same amount as the intermediate sample of quinua, and *chia* showed none at all. After waiting two days, I again shook the tubes, and a small amount of suds appeared in the *chia,* but still far less than in any of the others. Obviously more sophisticated tests should be carried out using a large number of samples before drawing any far-reaching conclusions.

in this way, but Wilson found plants deliberately cultivated for this purpose near Puebla. He also found that they differed slightly morphologically from the others, and he designated them as cultivar 'quelite,' using the local name for them. *Quelite* is a general term used for a large number of plants employed for greens in Mexico.

Going north from Mexico, we find that the chenopods were once widely used by Indians in the United States, both for their fruits and their greens. The wild species most commonly employed in the western states was *Chenopodium berlandieri*, which as we have seen was apparently the progenitor of the Mexican domesticate. In the eastern half of the country *C. bushianum* was most widely used. Although we have little archaeological information on quinua and none on the Mexican chenopods, there are excellent archaeological records for *C. bushianum*.

Chenopod fruits turn up in archaeological deposits in Illinois as early as 5000 B.C., but they do not become common until much later. By Middle Woodland times (ca. 150 B.C.) *Chenopodium bushianum* and other starchy seeds, knotweed and May-grass, become rather abundant, and David and Nancy Asch think they were all important food sources. Moreover, they feel that it is likely that all three plants were actually cultivated. We shall see in other chapters that there had been an increase in the fruit size of some plants in the archaeological record, and so we can be fairly certain that these plants were domesticated and had to be cultivated. The fruits of the chenopod, knotweed, and May-grass, however, do not differ significantly from those of plants of those species now found growing wild. This does not mean, of course, that they were not cultivated, for they could still have been intentionally grown and not differed from the wild plants, or they could have changed in characters other than the fruits, which have not been revealed by the archaeological material.

At present *Chenopodium bushianum* does not appear to occur in the abundance that some other species, such as *C. missouriense*, do. Why then, one might ask, did the pre-

historic inhabitants of the eastern United States collect
C. bushianum if these species had a similar abundance in
prehistoric times? *C. bushianum* has the larger fruits, but
people might easily have compensated for that by the ease
with which a greater number of fruits of *C. missouriense*
could be harvested. Could it be that the fruits of *C. bushianum* are more palatable? This seems to be a possibility,
for its fruits, as pointed out previously, appear to contain
little saponin. Of course, if the people were actually cultivating *C. bushianum,* they could have an ample supply of
fruits even if the species had a naturally patchy distribution.

Wilson has shown that *Chenopodium bushianum* and
C. berlandieri are very closely related. In fact, the former
differs from the latter only in its larger size, and the two
give fertile hybrids. Thus it seems apparent that *C. bushianum* is not really a distinct species but nothing more than
another variety of *C. berlandieri,* although the formal transfer of the name has not yet been made. Knowing this relationship, we can speculate more intelligently concerning
its origin. This "species," of course, could already have
been in the East when people started using it but perhaps
human beings were responsible for the introduction of
C. berlandieri to the eastern area, where it evolved into the
slightly larger *C. bushianum.* If so, then there is a parallel with the sunflower, which is thought to have been carried to the East as a weed before it was developed as a
domesticated plant. Still another possibility suggests itself.
Perhaps *C. bushianum* existed only as a cultivated plant,
and the populations of it today are simply escapes from
cultivation that have reverted to a weedy existence, still
retaining some of the larger size of the cultivated plant.

That there was also a domesticated chenopod in the eastern half of the United States seems clear from the presence
of large, light-colored fruits in archaeological deposits in
Kentucky and Arkansas. Wilson has examined an inflorescence from Arkansas which is of the compact type associated with domesticated plants, and he feels that it is similar
to *huauzontle.* Thus it must have come from Mexico. Dates

are not available for these specimens and they could well be quite late.

As with the sumpweeds (chapter 11), we have no definitely known historical observations of chenopod cultivation in the United States. There are, however, two accounts that some have thought may refer to a chenopod. The descriptions are so vague that positive identifications are impossible. Thomas Hariot, in *A Brief and True Report of the New Found Land of Virginia* (London, 1588), writes that Indians of eastern Carolina had "an hearbe which in Dutch is called *Melden.*" This name usually refers to a species of *Atriplex,* which is rather similar to a chenopod and a member of the same family. Le Page du Pratz in the *Histoire de la Louisiane* (Paris, 1758) mentions a plant grown by the Natchez Indians and calls it *belle dame sauvage. Belle dame* is a French name for *Atriplex.* Except for these possible occurrences, the cultivation of chenopods had fallen into disuse when the Europeans arrived, suffering the same fate, perhaps for the same reasons, as sumpweed.

Although earlier I indicated that the chenopods are likely to continue to be important food plants in the Andes and Mexico, I hardly think they will ever make a comeback in the United States, but I could be wrong. Presently there is an attempt to make quinua a crop plant in Colorado. Neither the Andean nor the Mexican chenopod is well adapted to most parts of the United States, but we still have *Chenopodium bushianum,* which grows well in the eastern United States. Maybe someone will want to see if it can be made a crop plant. Since it hybridizes readily with *huauzontle,* one might be able to use hybridization to attempt a rapid improvement.

Hybridization might also be employed to bring about improvement in the Andean and Mexican chenopods. Thus far the little improvement that has been done with them has been by selection—seeds from naturally occurring superior plants are used in an attempt to develop better plants. It is, of course, not unusual to find that crops of the Third World countries have not been subject to more plant breeding. These countries simply do not have the money or the trained people to carry out much experimental work. Moreover, it must be pointed out that the small flowers of chenopods make it difficult to secure hybrids. There is a cytoplasmic male sterile form of quinua that makes it easy to secure hybrids with it as the pistillate parent. However, the resulting hybrids will also be male (pollen) sterile, which is a definite disadvantage in certain types of breeding work. Yet if genes to restore pollen fertility could be found, it would make it possible to develop high-yielding hybrids, just as we do with corn. Simmonds, who developed the cytoplasmic male-sterile strain, did not encounter fertility restorers. However, Wilson in one cross did find fertility restored. Thus a hybrid quinua would be a distinct possibility. However, there probably is no place for it in the Andes. Quinua is mostly a peasant crop, and the people save their own seeds for planting. The hybrid seed would have to be produced anew by seed companies every year, and the peasants could not afford to buy them.

Through hybridization, however, the breeders might create improved strains that would breed true, and thus the farmer could save his own seeds for planting. Hybrids do not necessarily have to be made with the cytoplasmic male-sterile plant. Wilson found that by altering the day length he could produce plants that had pistillate flowers opening before the staminate ones. By using such plants with a marker-seedling recessive gene as the female parent and a pollen parent with the dominant gene, he obtained a number of progeny. Most of the plants were hybrids, and the few plants that resulted from self-pollination were readily recognized because of the presence of the recessive character and were discarded.

Both intraspecific and interspecific hybridization would seem to offer great possibilities for the improvement of quinua. Although the hybrid between quinua and *huauzontle* is nearly sterile, backcrossing is possible. Thus one might be able to transfer genes from the one species to

Otavalo Indians with quinua plants.

the other. For example, it might be feasible to produce quinua plants with less saponin by transferring genes to it from *chia*. Such breeding programs, however, are not carried out overnight, and they would require sustained financial support by governments or some other agencies.

Perhaps as one finishes this chapter he may have new respect, if not love, for the weeds in his garden. Although they may be nuisances today, some of them may have been the ancestors of the plants that now regularly grace our tables.

Sangorache growing in a tangerine orchard at Patate, Ecuador.

CHAPTER 5

Sangorache and the Day of the Dead

Soon after my arrival in Ecuador in 1962, I observed small gardens and orchards in many places around Quito. As was my custom, I paid particular attention to the weeds, and one plant I saw in a number of gardens was a pig-weed or amaranth that stood out because it was almost blood red in color. I thought little more of it at the time, but later, on returning to one of the gardens, I found that all the weeds had been removed except for the amaranth. This aroused my curiosity, and I soon learned that the amaranth was called *sangorache* and that it had a special role in the lives of Ecuadorans. The plant, insofar as I could learn, was never deliberately planted, but if it came up in a garden, it was allowed to remain. Thus it was an encouraged weed. Its well-developed inflorescence as well as its uniform color led me to believe that some deliberate human selection must have been involved in its ancestry.

Sangorache is used in Ecuador as a medicinal herb and as a coloring agent in the traditional beverage for Dia de Difuntos or Dia de Finados ("Day of the Dead"), simply called Finados by many Ecuadorans. This observance is thought by some to stem from the Ayamarca of the Incas, the month in which special tribute was paid to the dead, but because of its special significance in Ecuador I think it may well have been celebrated in Ecuador before the conquest of that country by the Incas in the fifteenth century.[1] Today it is a national religious holiday celebrated

[1] The Day of the Dead (*Dia de Muertos*) is widely celebrated in Latin American countries. It is particularly well known in Mexico, where, however, we would look to its origin from the Aztecs and Mayas. As in

101

on November 2, following All Soul's Day. On that day the Indians visit the cemeteries bringing food and beverages to the graves to nourish the souls of the dead, who are believed to return on that day to partake of the offerings. Nearly everyone in Ecuador, Indian and non-Indian alike, particularly in the highlands, eats the *pan* ("bread") *de Finados* and drinks *colada morada* or *muzamorra colada*, a thick corn soup colored purple. It is in the latter that inflorescences of sangorache are used as a coloring agent. Ever since learning about sangorache in 1962, I have wanted to know more about Finados and *colada morada*, and in 1982 I had the opportunity to do so with the help of Sergio Soria and his family.

A week before Finados, the *pan* or *huahuas* (a Quechuan word for children or babies) makes its appearance in the markets. The ones I saw were in the form of human figures (similar to our gingerbread men), llamas, or doves. Some were decorated with icing. On the Friday (October 29) before Finados I visited markets in Quito. Inside the larger markets I observed large amounts of sangorache for sale, and on the block outside the central market numerous vendors were selling bundles of herbs used in flavoring *colada morada;* these included branches of *arrayán* (*Eugenia* sp.) and orange tree, *yerba luisa* (lemon-grass), and *cedron* (*Lippia citrodora*) in addition to inflorescences of sangorache. A packet cost five *sucres*, or about ten cents.

On Saturday we went to Ambato, a fair-sized city about two and one-half hours drive south of Quito, and I soon found that Finados was hardly less important than Christmas — there, at least. About twelve blocks of the widest

---

Ecuador, the observance in Mexico involves an intermingling of Catholic and Indian religious customs. Some of the customs in Mexico — for example, offering food to the dead — are similar to those in South America. Some of those practices may well have antedated the Incas, Aztecs, and Mayas in the Americas.

Timothy Johns (letter, June 18, 1983) informs me that an amaranth is also used in All Saint's Day in parts of Bolivia, but it is not the same species as that in Ecuador, and it is used in a different way.

Pan de Finados in the market in Ambato.

street had been blocked off to cars, and stands were being set up in the middle and along the sides offering a great variety of wares, mostly ceramics, none of which, however, seemed to have much to do with Finados. Many of the vendors had come from some distance. On other streets stalls were provided with paper *coronas*—wreaths—for the decoration of the graves and the flower market was more filled than usual. In particular there was an abundance of *illusiones (Gypsophila paniculata)*, which we know as baby's breath, a favorite flower of the Indians for decorating the graves. What special significance it has for the occasion I do not know, and someone suggested that it was preferred by the Indians because it was one of the cheapest flowers available. In the food markets, in addition to large amounts of *pan de Finados,* there were large stacks of sangorache, *arrayán,* orange branches, and *yerba luisa:* These were sold separately instead of in packets as in Quito. *Mora* (blackberries) and *mortiño* (a native blueberry) were in greater supply than usual. In nearly all of the little grain stores near the market I saw the ground kernels of *maiz negro* ("black corn"). A week earlier I had observed a large number of ears of *maiz negro* in both Ambato and the nearby town of Pelileo. This was of some interest to me, for I had never seen it before, and on my first trip to Ecuador I had heard that it was very rare. In fact, there is no mention of it in the comprehensive *Races of Maize of Ecuador* by David Timothy.

I also observed numerous vendors of fruit juices selling *colada morada,* and the next day I decided to learn more about how it was made. The most detailed recipe was provided by Syla Cepeda. In one bowl she boils sangorache and brown sugar along with the "aromatics"—*arrayán, yerba luisa,* leaves from the orange tree, the rind of a pineapple, *albahada* (basil), and *ishpingo* (*Acotea quijos*). In another bowl she mixes blackberries and a pineapple, which have been put through a blender with three strained naranjillas. The contents of these two bowls are added to flour of *maiz negro* and boiling water. Pieces of *babaco* (*Carica pentagona*) and strawberries are added before drinking. In talking with others I learned that there was considerable variation in the ingredients used—some people use fewer or other aromatics. Some recipes call for pimientos (sweet pepper, *Capsicum annuum*), and some people use *mortiño* in place of, or in addition to, *mora.* Whether pieces of fruit are added, and the kinds of fruits, seems to be a matter of individual choice. One thing is constant—the rich purple color of the final product.

Later in the day I had my first taste of *colada morada* and *pan de Finados* at the Sorias'. The *colada morada* was usually served warm—they told me that was the traditional way— but that night we drank it cold, the way they preferred it. Either way, I found it to be a most delicious, satisfying drink. They told me they didn't like the taste of *maiz negro,* so they simply used corn starch, and I learned later that in parts of the country where *maiz negro* is not found, flour from other corn is used.

Maiz negro and yellow corn for sale at the market in Pelileo.

Although the *colada morada* that I saw was nonalcoholic, there seems to be general agreement that the corn was once fermented to make a corn beer or *chicha,* and some Indians may still follow this practice for making *colada morada.* When it was first used and the original recipe of course are unknown. Most of the ingredients used today are from native South American plants, but some of them, such as leaves of the orange tree, came from plants introduced after the Spanish Conquest. I suspect that corn has always been the most basic ingredient. Perhaps sangorache was first used when some people found that they could not get the *maiz negro.* Certainly the color of *colada morada* must be important. Four plants contribute to it—*maiz negro, mora, mortiño,* and sangorache—although by itself sangorache yields a deep red color. The people whom I asked could not tell me why it had to be purple, but I suspected, even though the color was purple instead of red, that it might represent blood—the blood of the ancestors. Thus the drinking of *colada morada* and the eating of *pan de*

Finados would be a communion with the dead, perhaps borrowed from the Catholic religion after the Conquest. An Ecuadoran anthropologist, Guevara, however, offers another explanation: the *colada morada* does symbolize blood, but it is the blood of llamas or human beings who were sacrificed in the ancient religious practices of the Indians. Today, however, most people drink *colada morada* as an enjoyable beverage and as part of the celebration of Finados with no knowledge of how or why the practice originated. It is now simply a custom.

On All Soul's Day we went to the cemetery in Ambato, and it was the scene of great activity; people were white-washing the stones, replacing the lettering on the graves, and leaving decorations of wreaths and flowers. The next morning, the day of Finados, we visited the cemetery at Pelileo, and it was even more crowded with people. In the Indian section of the graveyard there were small groups of Salasaca Indians around individual graves making offerings of food and drink to the dead just as they had done for centuries. Some of them had bread in their hands, and there were small dishes on the ground. I didn't go close enough to identify the contents. There were also wet places on the ground, and from the bottles in evidence I decided they must be from some alcoholic beverage, probably *aguardiente* (which is made from sugar cane) or wine. In one place I saw a whole cigarette had been left, and candles were present. Beside one small stone I saw a small bowl of milk; no doubt it was an infant's grave.

Since my interest in Dia de Difuntos stemmed from my acquaintance with sangorache, more about the plant is in order. After my first visit in 1962 I reported two other names for the plant, *ataco* and *quinua de Castilla,* but on subsequent visits I learned that at least one of these names, if not both, refers to a different amaranth. *Quinua de Castilla* is very similar to sangorache, particularly in its well-developed inflorescence, but I found that I could distinguish the two by color, *quinua de Castilla* being more brownish than red. To which plant the name *ataco* was originally

Salasaca Indians making offerings to the dead in the cemetery at Pelileo.

applied, I am not sure; today people use it for both of these plants.

In my paper on sangorache I identified it as *Amaranthus quitensis*. Since that time M. Patricia Coons has made a detailed study of the amaranths of Ecuador, and she concluded that *A. quitensis* is a synonym of *A. hybridus*, a wide-ranging species that is common in North America as well as South America. Although she considered that sangorache is sufficiently distinct to be recognized as a variety, she has not yet published the varietal name. From her study of the flowers it became apparent that, in spite of its external resemblance to sangorache, *quinua de Castilla* belonged to a different species, *A. caudatus*. Quinua de Castilla, like sangorache, seems to be an encouraged weed, but I found it in an Indian's garden on the campus of the Universidad Central in Quito, where it was definitely cared for, if not intentionally planted.

Both sangorache and *quinua de Castilla* are found regularly in the medicinal section of the larger markets in Quito. I learned from the vendors that they are used in the treatment of a variety of ailments, but the answers I was given were not always consistent. Several friends, however, told me that sangorache as a tea or infusion in water is widely used for kidney and liver ailments. In his book on medicinal plants of Ecuador, Alan White gives "ataco, sangorache," which he identifies as *Amaranthus caudatus*, and he states that it is used for throat irritations, diarrhea, hemorrhage from the bowels, and excessive menstruation. It is not clear from his account, however, whether these are uses for the plant in Ecuador or for amaranths in general. He does not include *quinua de Castilla* in his book.

The names *sangorache* and *ataco* are of Indian origin, but the name *quinua de Castilla* is a hybrid from quinua, the native pseudocereal, and Castilla, an ancient kingdom now a part of Spain. The resemblance of amaranths to chenopods can perhaps explain the first part of the name, but *Amaranthus caudatus* did not come from Castile. The name does suggest, however, that *A. caudatus* may be an introduced plant.

Amaranths, like chenopods, were ancient food plants in the Americas, two species having been domesticated in Mexico and Central America and one, *Amaranthus caudatus*,[2] in the Andes. The domesticates have pale-colored seeds in contrast to the dark seeds of sangorache, *quinua de Castilla*, and the wild species. The domesticated form of *A. caudatus* is still grown in Peru as a grain crop, but it appears to be very rare in Ecuador. The only place I have seen it is near Cuenca, where it was called quinua. The only possible reference that I have found to the plant in Ecuador is in Cordero's work. He refers to a *quinua de tostar*, the small seeds of which are eaten and which he identifies as *"Amaranthus?"* Perhaps *quinua de Castilla* is a stabilized hybrid derivative of the domesticated *A. caudatus*

---

[2]*A. candatus* is sometimes also grown as an ornamental in the United States under the name love-lies-bleeding.

Quinua de Castilla. (Courtesy M. Patricia Coons)

and sangorache. Coons's study has shown that hybridization between the two species may occur.

In recent years there has been an attempt to develop amaranths as food plants in the United States. The movement has been promoted by Robert Rodale, the publisher of *Organic Gardening,* and a number of farmers are now growing them on a small scale, chiefly as a leafy vegetable. Because seeds of amaranths have substantial amounts of protein high in the amino acid lysine, the plants certainly deserve more use as a grain crop. It is difficult to predict how successful the crop might become in the United States. Amaranths, however, could well contribute more to the diets of the Latin American Indians than they presently do. Possibly plant breeding through hybridization could lead to strains superior to those now being grown. No great future can be predicted for sangorache unless it were to contribute genes for the improvement of the grain amaranths. Fortunately, it appears in no imminent danger of extinction. Weeds are tough plants, and as gardeners know, they are difficult to eliminate.

CHAPTER 6

A Trip to Tulcán

It seems fitting that a chapter involving a visit to a cemetery be followed by another on the same subject. It will also provide something of an interlude. One of my most memorable trips around Ecuador with Jaime Díaz in 1962 was to Tulcán, which is on the Colombian border.[1] Before reaching the city, we passed through the *páramo* of El Angel, and the vegetation there has to be the most remarkable that I have ever seen.

Early visitors from Europe referred to the *páramos*, the cold, often fog-shrouded zones above the timberline, with their distinctive plants, as moorlands or grasslands, and to the higher ones, where all vegetation becomes scarce, as deserts. Frederick Hassaurek, who was the minister of the United States to Ecuador and arrived there in 1861, has left us a good account of *páramos* in his *Four Years Among Spanish Americans* (1881):

Paramos, properly speaking, are the highest plains or heaths of the Cordillera, covered with high tufts of long and dry grass, which the natives call *"paja"* (straw). . . . Its aspect is dreary and cheerless in the extreme. But paramos may also prove dangerous to travellers. Winds laden with icy vapors blow over them with tremendous violence, when as the natives call it, the paramo *"se pone bravo"* (gets wild). Dense fogs frequently envelop man and beast; darkness covers the earth and conceals every trace of the

---

[1] Another particularly memorable trip, but for another reason, was to the port of Esmeraldas. There was some concern at the time that pro-Castro Cubans might attempt to infiltrate the country, and as I was obviously a foreigner and was not carrying my passport at the time, I ended up in jail. Jaime had to visit a series of officials before securing my release.

The *páramos* of El Angel from a distance.

road; snow, hail, or sleet comes down unmercifully; and often the traveller loses his way and wanders helplessly over endless heaths. But this is not the worst; when worn out with fatigue and hunger, benumbed with cold and unable to urge on his jaded mules, he dismounts and sits down to recover his exhausted strength, his stomach soon becomes affected as if at sea; his blood ceases to circulate, his muscles grow stiff, and he expires with a ghastly smile upon his features. Travellers thus found dead in these inhospitable regions, are said to be *emparamados.*

Today good roads and the automobile make travel considerably easier than it was in Hassaurek's time, but *páramos,* can still be desolate places. Although grasses are the dominant vegetation, other plants are found in them.

I have visited many *páramos,* but to my way of thinking none can equal El Angel. As we approached it, I could see thousands of tiny dots on the landscape, and as we got nearer I could make out that each dot was an individual plant, known in Ecuador as *frailejón* ("large friar"). Rather than attempt a description, I refer the readers to the first

As one gets nearer, the *frailejones* become more apparent.

three photographs illustrating this chapter, and they can decide for themselves if the plant deserves its name. The *frailejón, Espeletia hartwegiana,* is a member of the Composite family. It is the only member of the genus represented in Ecuador, but a number of other species of the genus occur in both Colombia and Venezuela.

After we had returned to the jeep, I asked Jaime about the principal occupation of the people in Tulcán. He replied that 45 percent of the people were engaged in smuggling Ecuadoran products into Colombia, and that an equal number were involved in smuggling Colombian goods, chiefly clothing, into Ecuador. Thinking quickly, I asked about the other 10 percent, and without hesitation he answered that they were too lazy to work. He was exaggerating, of course, but a great deal of smuggling did occur between the two countries and still does, I understand.

As we entered the town, Jaime told me that there was only one thing worth seeing in Tulcán, and that was the

Ing. Jaime Díaz and his assistant examining flower heads of *frailejón*.

cemetery. That sounded strange to me, but it was to prove to be correct. Tulcán itself was a rather dismal place and quite cold, but the trip to the cemetery the next day more than made up for the deficiencies of the town. Nearly all of the trees and shrubs in the cemetery, mostly cypress, were trimmed into a great variety of ornamental shapes, a kind of gardening known as topiary. Although ordinarily I prefer a shrub or tree to look like a shrub or tree, I was impressed by this work. I was all the more impressed when I was told that the man responsible for the sculpturing was self-taught in the art, having been inspired by photographs he had seen in books. The practice of topiary apparently originated with the Romans and still may be seen in many places in Europe, as well as a few in the United States. When Dr. Edgar Anderson, who was an authority on topiary, saw my photographs, he said that these were the best examples he had seen of topiary work in the Americas. Rand McNally's *1983 Handbook of South*

Topiary masterpieces in the cemetery at Tulcán, Ecuador, 1962.

America, however, is even more enthusiastic, stating that the "topiarist's art reaches its apogee" in Tulcán.

In my trip of 1982 I had hoped to revisit Tulcán, but because of a transportation strike I was unable to do so. I had wanted particularly to learn the artist's name, to learn if the topiary was still maintained there, and to identify the species used. Friends told me that it was still as beautiful as ever and that the same man was still in charge and was training one of his sons to take over after his death.

After having reread what I had written above, I was far from satisfied with it. I resolved that I had to make another trip to Ecuador, and it was a good thing, for some of the information that I had been given previously was not correct. In June, 1983, my wife and I landed in Quito, and a couple of days later we were on our way to Tulcán by bus. The trip took a little over four hours, half of what it had

Another selection of topiary in the cemetery at Tulcán, Ecuador, in 1962.

taken earlier, for the road is now much improved, though unfortunately the new route does not go by way of El Angel and we did not get to see the *páramo*. We arrived late in the day and were soon settled in Rumichaca, a government tourist hotel right on the border with Colombia, some five kilometers from the city itself. Because it was so late, we decided to postpone our visit to the cemetery until the next day.

The next day was a beautiful one, and Tulcán, which had grown considerably since my earlier visit, was not nearly as dismal as I had remembered it. We spent about two hours inspecting the cemetery. There had been some changes in the faces on the shrubs lining the main walk, but overall the topiary was even better than I had remembered.

At the hotel I had been told that the artist's name was Azael Franco. I had particularly wanted to talk to him, but as it was a Sunday morning, he was not at the cemetery.

Same view as the previous one, in 1983. Note that shrubs on the right are beginning to need trimming.

We asked people at the gate house for his address. They did not know if, so we consulted a telephone book at a nearby store. A street was given, but no house number. We walked to the street, and following a set of conflicting directions from various people, we finally found his house.

A girl let us in and took us to the living room. On the wall were several awards that Sr. Franco had received, and from them I learned that his full name was José Maria Azael Franco Guerrero. After a couple of minutes he entered the room and seemed delighted to see us and not at all surprised to have visitors from the United States. I presented him with an article, with his photograph, which I had written about the cemetery twenty-one years earlier. One of his sons, who now helps with the work at the cemetery, soon joined us. Although it was only 11:30 in the morning, we were served drinks of scotch, a tradition for visitors in Ecuador at any time of the day.

Sr. Franco, who was born in the town of El Angel, was

Topiary in the cemetery at Tulcán in 1983.

Dorothy Heiser admiring topiary (1983).

Mother and child. The masterpiece of 1983.

The topiary artist Sr. Franco in the patio of his home, 1983.

now eighty-four years old and still worked at the cemetery. The city provided the labor for the actual trimming of the shrubs. The topiary was begun in 1936, when the city officials hired Sr. Franco and he decided that "something grand" should be done for Tulcán. I was not quite correct earlier when I said that he had had no training, for I learned he had earlier practiced horticulture under the direction of Don José Tamayo, who had "specialized" in the United States. The cypress in the cemetery had been identified as *Cupressus sempervirens* (= evergreen),[2] a species native to southern Europe. The original plants are still there today, and Sr. Franco told me that he expected them to last five hundred years, which would not be surprising, for many gymnosperms will live to a great age.

---

[2] I wanted to verify the identification, but I could not find any cones. One identifies flowering plants by their fruits (or the flowers), and one usually needs the cones to identify gymnosperms, for they have no flowers and consequently no fruits.

CHAPTER 7

Chochos and Other Lupines

The first plant I saw on my arrival in Quito in 1962 was a lupine (*Lupinus pubescens*) growing naturally on the grounds in front of the airport. This handsome species is no longer found there, but it does still occur on vacant lots and roadsides around the city. Although *Lupinus pubescens* is probably the most common species in Ecuador, it is by no means the only one or the most striking.

The genus *Lupinus* has centers in North America, South America, and the Mediterranean region. The Texas bluebonnet, the state flower, is one of the best known species of the United States. Some eighty-five species are reported for the Andes. The genus has furnished us with both food plants and ornamentals for the garden.

To gain some idea of the diversity of species in Ecuador, one can hike from Quito at 9,370 feet up through the *páramos* to the peak of volcano Pichincha at over 14,500 feet. It was on these slopes that General Sucre defeated the Spanish to gain Ecuador's independence in 1822.

On my trip to Pichincha, Ecuador, in 1962, I was accompanied by Saulo Soria, the younger brother of Jorge, my former student. At first we passed through groves of *Eucalyptus,* about the only tree of any size in the sierra today and a very important source of wood both for fuel and other purposes. Any time I smell *Eucalyptus* today, I think of Ecuador. The tree is not native there, however, although it has the appearance of being so in some places; it was introduced from Australia less than one hundred years ago.

As we continued upward we passed near the huts of Indians who cultivated small fields and pastured cattle at still higher altitudes. On the path we met a little girl, half running and carrying two pails of water. The sight inspired both admiration and envy, for at that altitude every step was becoming more difficult for me.

Lupinus pubescens stayed with us for the first three thousand feet, but other species, some rather similar to it, soon began to appear. Just before we reached the *páramo,* we saw a very different one, *L. microphyllus.* The species epithet translates as "small-leaved," a very good name for it, for the leaves are only one-tenth the size of those in most of the other species. It grows nearly prostrate, and one almost has to get on his belly to see it. Such "belly plants" are common on the *páramos.*

Finally, at 14,500 feet, where we found patches of snow, two other species occurred and the sight of one of them, *Lupinus alopecuroides,* made us temporarily forget the scarcity of oxygen at that altitude. For a description of this plant, I return to my friend Professor Jorge Tinajero, although my translation can hardly preserve the eloquence and enthusiasm of the original Spanish: "No plant in all the world equals that of *Lupinus alopecuroides,* whose inflorescence extends a yard and a half and is more than six inches in diameter, and is entirely covered with a whitish velvet, studded here and there with little flowers of an intense blue. . . . it bears the name 'Tail-of-the-Fox,' when in reality it is a fur of the ermine, sprinkled with amethysts." Although I have never seen one quite that size, nor do I entirely agree that it is without an equal, it nevertheless is a most impressive plant. One other lupine, *L. nubigenus,* although not as large as tail-of-the-fox, has an equally dense pubescence and grows at still higher elevations. Unfortunately, I never encountered it on any of my trips.

After lunch and a much-needed rest, we started back down, taking a different route. The descent was fairly rapid, and we were soon going by the fields of the Indians again. This time, among the plants we saw in cultivation was a lupine (*Lupinus mutabilis*) generally called *chocho,* the

Spanish name, in Ecuador and *tarvi,* an Indian name, in
Peru and Bolivia. This certainly has to be one of the most
attractive of all domesticated food plants.

The chocho is an annual, two to four feet tall, with
smooth leaves and stems, in contrast to the hairy ones of
most of the wild species (which incidentally are called
ashpa chocho, using the Quechuan word which, loosely
translated, means "wild" or "of the field or woods"). The
flowers of the chocho are nearly an inch in diameter and
are usually some shade of blue touched with white and
gold. The chocho is widely grown in the highlands from
about six thousand feet to nearly twelve thousand feet and
does well in poor, dry soils. Occasionally one sees a whole
field devoted exclusively to chochos, but more commonly
they are mixed in fields of quinua and dwarf maize. Sometimes they are planted around the edge of a field of maize,
where, according to some people, they serve as a living
fence, since they are unpalatable to cattle. In fact, several
species of lupine in the western United States are known
to be poisonous to livestock. Although they are not eaten
by cattle in Ecuador, there is no assurance that the cattle
will not go through them to get to the maize.

Since the pods of chocho, slightly larger than those of
most lima beans, are indehiscent, the seeds are retained
on the plant, in contrast to the splitting pods of the wild
species. Each pod contains two to five oval seeds, several
times the size of those of the wild species. The seeds in
Ecuador are generally white, but black seeds and black and
white mottled seeds are known. The seeds contain alkaloids
and cannot be eaten directly. One wonders, as with other
plants so poisonous that they require special preparation
before being used as food, how people ever learned to
prepare them for human use. In some way they found that
the seeds were edible after being washed in water, which
leaches out the alkaloids. The method in use in Ecuador
today is to soak the seeds for several hours, then, after
cooking them, to place them in running water, usually in
a stream or river, for several days. They are then sold

Flower stalks (about 20 inches tall) of tail-of-the-fox on the *páramos* of Pichincha.

in all principal markets and on street corners in Quito and other cities in the highlands. The Indians often buy a handful for a few centavos and munch on them the way we would peanuts. The taste is rather difficult to describe, but it is more like that of a nut than that of a bean. The seeds are also added to various cooked dishes, and diced they are a common ingredient in the pepper sauce that is found on almost every table in the country.

The chochos, however, should be regarded as more than a condiment or a between-meals snack, for they are extremely rich in protein and oil and thus supply important additions to the starchy diet of the Indians. Meat is a luxury to the Indians of Ecuador, and this is one of the reasons that the chocho retains some importance in that country. Much of the land that once may have been devoted to its cultivation, however, is now given over to other legumes, particularly broad beans and peas, introductions from Europe, which grow well in the highlands. These plants were more acceptable to the Spanish conquerors, and also their ease of preparation for table use helps explain why they tended to replace the chocho.

A few years ago concern was expressed that the chocho would disappear and that a "cultural heritage" would be lost, as Carl Sauer expressed it. George Carter wrote that its loss would be the "wastage of a basic resource," and although the plant might not be particularly valuable at present, a time might come when it would be more useful. As a plant that is an excellent source of protein, is rich in oil, is relatively disease-resistant, and grows well on poor soil, it would, one would think, be worth preserving, quite apart from the fact that the loss of any creation of nature and human beings, as the chocho is, would be a sad event.[1] Had it disappeared, the responsibility would not rest with the Indians who grew it, for they must grow the plants that give them the greatest returns, but with the governments that failed to encourage the development of the plant. Fortunately, with an awakening of interest in some of the Indians' crops in recent years, some attention is being paid to the chocho. The Peruvian government now has an active program at Cuzco for its improvement. A living collection of plants from many places in the Andes is being maintained there. Two Germans, Erik von Baer, working in Chile, and Rainer Gross, working in Peru, have carried out a number of experiments, and through selection they have been able to secure strains in which the alkaloid content of the seed has been reduced tenfold. They have also worked toward increasing the oil content of the seeds.

[1] The chocho also improves soil fertility. It is a member of the legume family and, like most members of the family, has a symbiotic relation with certain bacteria which live in nodules on its roots. These bacteria are able to fix atmospheric nitrogen. Nitrogen is a constituent of protein, which helps explain why the plant is so rich in that nutrient.

Chochos and *capulís* (a black cherry) for sale on a street corner in Quito.

Even though the future of the plant still seems uncertain, its demise certainly does not appear imminent.

Thus far I have found little in the early literature about the chocho. Padre Bernabé Cobo, who in the seventeenth century gave us an excellent account of many of the culti-vated plants of the New World, fails to say much about the chocho except that there was an abundance of wild *altra-muzes,* which the Indians called *tar-ui,* in the fields, and that they grew to such heights they served for fuel. Gar-cilaso de la Vega recorded that his mother's people had "lupins like those in Spain, only rather larger and more white, which they call *tarvi."*

The origin of the chocho has thus far received little attention. In fact, the only article that I have found deal-ing with the subject is by two Polish plant breeders, T. Kaz-imierski and E. Nowacki, in 1961. Their rather unusual

Flowers and leaves of chocho *(left)*, a wild lupine, *Lupinus pubescens* *(right)*, and a natural hybrid between them from near Cayambe, Ecuador.

hypothesis supposes that two North American species, *Lupinus douglasii* and *L. ornatus*, were carried to the Andes during southern migrations of the Indians, and there, through hybridization with *L. pubescens*, they somehow gave rise to the cultivated species, *L. mutabilis*. So far as I am aware the North American Indians did not use seeds of any lupines for food, and before such an elaborate theory for the origin of the chocho is accepted, other possible explanations need to be explored. Although I found no wild species of lupine in Ecuador that were similar to the chocho, I did find, in going through herbarium material from Peru and Bolivia, several species, apparently wild, that appear to be close to it. One of these is *L. montanus*, and I think we should look for the chocho's origin in highland Peru or Bolivia.

As was pointed out earlier, the genus *Lupinus* also is

A wild lupine (*Lupinus paniculatus*) near Baños, Ecuador.

well represented in the Mediterranean region. There, in prehistoric times, three species were brought into domestication: the white lupine (*Lupinus albus*), the blue lupine (*L. angustifolius*), and the European yellow lupine (*L. luteus*). The genus therefore is one of the very few that has contributed plants for human use in both the Old World and the New. Sweet varieties, forms that contain little or none of the bitter alkaloids, have been developed by plant breeders in this century for these species. They are grown mostly in eastern Europe, where they are principally used for animal feed. One of the Old World species, *L. albus,* is marketed in this country under the name "lupini beans" for human consumption. The seeds require considerable soaking, however, before being cooked, but lupini beans, in jars, ready to eat, are available in Italian markets in the larger cities of our country. In appearance and taste they are rather similar to the chocho.

CHAPTER 8

Green "Tomatoes" and Purple "Cucumbers"

By this time readers must think that I deal mainly with little-known plants. They are correct. I have given that impression intentionally. Whereas most of our major crops such as corn and potatoes, have attracted a lot of attention from scientists, and justifiably so, many of the minor ones have received scant notice. Some of those minor crops, I feel, may reveal a great deal about the process of domestication, perhaps more readily than those more extensively used by human beings. Thus they deserve study.

Although numerous plants now used as food in the United States come from many parts of the world, there are still many food plants that are completely unknown or relatively rare in this country. The *tomate* or *tomatillo,* or green "tomato" of the title, belongs to the latter group, although it is becoming known in our grocery stores as more people discover the joys of eating Mexican foods. The *pepino,* or purple "cucumber," is presently little known in the United States. Both of these plants were treated in my book on nightshades some years ago, but I now know a lot more about them than I did at that time. Much of this knowledge comes from the work of two of my former students, Gregory J. Anderson and W. Donald Hudson, who studied them for their Ph.D. dissertations.

Plants like the tomate, belonging to the genus *Physalis* (family Solanaceae), are annual or perennial herbs. Commonly they are called husk tomatoes or ground cherries, common names that are somewhat more appropriate than most. The calyx enlarges with the fruit so that it com-

Uvillas (cape gooseberries) hanging on the wall, and pepinos, below right, in a market stall in Quito, Ecuador.

pletely covers the fruit like a husk at maturity, and the fruits are often the size of cherries and are borne near the ground. There are several wild species in the United States and some people enjoy collecting the rather pleasant-tasting fruits of several of them. Many seed catalogs include some of the species among their vegetables, and people grow them to make the fruits into preserves or pickles or for eating raw. In the same catalogs, one species, *Physalis alkekangi,* the Chinese lantern plant, is offered among the ornamentals and is grown for its showy orange or red calyx. The color persists as the plant dries, and they are used in winter bouquets. The Chinese lantern plant, unlike the others to be discussed here, is native to the Old World. It was known to Dioscorides, who ascribed medicinal virtues to it. The species epithet is from an Arabian name.

From the Andes comes *Physalis peruviana,* and although some sources state that it is, or was, cultivated there, I have found no evidence of its cultivation today. I have seen it in many places in the Andes as a weed or in gardens, where it comes up spontaneously. It appears to be an encouraged weed rather than a plant that is deliberately sown. The sweet fruits, usually yellow at maturity, are eaten raw. Occasionally one sees bunches of the fruits in markets in Ecuador, where the plant is known as *uvilla* (Spanish for "grape"), and in other countries it is called *uchaba* or *capulí.* Although apparently not a cultivated plant in the Andes, it is planted in other parts of the world today, but the few of these plants that I have seen differ little from those of the Andes. I have not been able to document the travels of the plant with certainty, but apparently it was carried to the Cape Province of South Africa, where it became cultivated, and from there it was taken to New South Wales, where it became known as the cape gooseberry, the English name most widely used today. That name is about as logical as Jerusalem artichoke or Irish potato, but then, as pointed out elsewhere, logic does not always prevail in the selection of common names. The plant itself is rather similar to our native husk tomato, *P. pubescens,* but the latter is a diploid, whereas *P. peruviana* is a tetraploid. No studies have been carried out in an attempt to determine what diploid species was or were involved in its origin.

In the United States today some seed houses offer the native *Physalis pubescens,* sometimes identified as *P. pruinosa,* which also appears to be very recent as a cultivated plant. The Indians of what is now the United States probably ate the berries of species of *Physalis,* but there is no record of their having cultivated any of the species.

It is in Mexico and Guatemala where one species has long been cultivated. In most works its name is given as *Physalis ixocarpa,*[1] but it should correctly be called *P. phil-*

---

[1] *Ixocarpa* means "with viscid or glutinous fruits." The young fruits of this species, and several other species as well, are quite sticky.

Uvillas, or cape gooseberries, in the market in Quito. The husks have been removed from the berries in the small plate. The berries with the husks are below them.

adelphica (the wild form of this species is found in the eastern United States). One sees the green fruits, called tomates, in markets throughout much of Mexico and Guatemala and increasingly in many places in the United States, where they are usually called *tomatillos.* The fruit is employed in the making of the green sauces used on eggs, meat, or rice and is an essential ingredient in *enchiladas verdes* ("green"). The sauce is made by mashing the fruits, adding onion or garlic, chili pepper (usually serrano, but milder sorts may be used if one prefers), coriander, salt, and pepper, and then cooking the mixture. Various other ingredients are sometimes used. Recipes can be found in most Mexican cookbooks. Cooks no longer have to make their own green sauce, however, for bottles of it are now available in many stores in the United States. It is often labeled taco sauce but is green instead of red. To balance

Tomate in flower.

my unkind remarks about the taste of some of the foreign food plants, I should point out that I find the green sauce a welcome addition to many foods. The fruits are seldom used alone or eaten raw, for they are rather bland—L. H. Bailey once described them as "mawkish." The taste is a little like that of a "real" green tomato.

Here I had better distinguish between tomates and tomatoes. The word *tomate* (from the Nahuatl *tomatl*) is employed for a number of solanaceous fruits in Mexico, with modifiers to distinguish the different kinds of plants. The word alone is used for the domesticated *Physalis,* and the wild plants of the genus are called *miltomate,* probably meaning "tomate of the fields." The tomato (*Lycopersicon esculentum*) became *jitomate* in Mexico, but when the plant went to Europe the modifier was dropped, and it became *tomate,* which was later converted to *tomato* in English. In

Tomates in the market at Morelia, Mexico. Wild type (*left*); domesticated (*right*). The husk (calyx) is still present on the berries. (Courtesy of W. D. Hudson).

other parts of Latin America *tomate* is also used for *Lycopersicon esculentum*.

Although all of its wild relatives are native to western South America, the tomato apparently became a domesticated plant in Mexico. Some years ago, James Jenkins, a student of the tomato, concluded that its wild progenitor, a plant very much like the cherry tomato of our salad bars, in prehistoric times was introduced to Mexico, possibly by birds, where in time it became cultivated and eventually domesticated. In fact, Jenkins suggests that its similarity to *Physalis* may have led to its domestication, although I fail to see much similarity. Unfortunately, the archaeological record so far has been of no help in unraveling the times and places for the origins of these plants. Of course, it is unlikely that the fleshy fruits would be preserved archaeologically, but the seeds might. Thus far there are no reports of tomato seeds. Seeds of *Physalis* have been found at Tehuacán, Mexico, but it is not known if these are from domesticated or wild plants.

In visiting markets in Mexico, I observed that often there were two kinds of tomates for sale: small ones, less than an inch in diameter, sometimes called *miltomate*, and large ones, frequently two inches in diameter. The large ones I knew came from the domesticated plants, and Hudson subsequently showed that the small ones came from wild or weedy plants which grow spontaneously in many places, although at times they may also be cultivated. The small and large fruits come from the same species, and without much doubt the domesticated plants were originally derived from plants similar to the spontaneous ones. Somewhat to my surprise I found that the cost of a kilo of the smaller ones was greater than that of the larger, domesticated ones. Inquiry among my Mexican friends revealed that many people prefer the smaller ones because they have a better flavor, and later the well-known Mexican botanist Efraim Hernandez X. informed me that the smaller ones blend better with chilis to make sauces. The only difference that Hudson and I could find is that the smaller ones at times seemed to be a little sweeter. This brings up a question: Why have the domesticated variety at all if the wild ones are preferred? I am not sure of the answer, but I imagine that the domesticated one outyields the wild type and is necessary to fill the demand for the fruits. It also may be easier to grow for reasons that I shall mention shortly.

One may also ask why, if the wild variety is so desirable, did people ever develop a domesticated variety with large fruits? We can assume, I think, that after wild types were originally cultivated, mutants appeared having larger fruits, that seeds from these were then grown, and eventually a type having large fruits became the commonly cultivated type. Could this have been done because a farmer had a certain pride in producing larger fruits than his neighbor and that somehow "bigger became better" for that reason? On the other hand, perhaps it was not the

Berries of tomate. Domesticated (*left*); wild type (*right*).

size of the fruits that was being selected but some other
character associated with larger fruits. Hudson has shown
that seed size is correlated with the size of the fruits. Not
only do the large fruits have larger seeds, but those seeds
germinate more readily and more evenly than seeds from
the smaller fruits. Moreover, as would be expected, the
larger seeds produce larger seedlings. Thus if a mixture
of large and small seeds were planted, the seedlings from
the large seeds might have an advantage. If the farmer
did not eliminate the smaller, later-germinating seedlings
by weeding, they might well be crowded out by the larger
ones. Thus the larger-fruited plants might in time become
more common in cultivation than the small-fruited sorts
and eventually become the only type in cultivation. This
development did not necessarily have to result from intentional selection for larger fruits as such, but for the advan-
tages conferred by larger seeds. In other words, if there is

a link between fruit size and larger, more rapidly and evenly germinating seeds, the people could not help but produce larger fruits. One may ask if this did not occur in other domesticated plants as well.

As I observed the tomate in many markets in Mexico, I was surprised to find that, except for the size, the fruits showed little variability. This lack of variability is rather striking when the tomate is compared to many other domesticated plants, particularly the chili peppers that were seen in the same markets. Such a lack of variability might characterize a recently domesticated plant, but although we do not know how old the tomate is, it can hardly be considered a recent domesticate. Perhaps the nature of the fruit did not allow for the great variability that is found in the chili peppers, but then the tomate fruit is similar to the tomato in being a fleshy berry, and yet the tomato fruit shows far more variability in color, shape, and size. Neither of these possibilities seemed to offer a good answer, but the explanation may be found in the pollination system. The tomate was shown to be self-incompatible by K. Pandey some years ago, and this finding was verified by Hudson for a large number of accessions. This means that the flowers will not self-pollinate but that the pollen must come from another plant. Many of the domesticated plants that show great variation are self-compatible, thus allowing self-pollination to take place; thus when any variant arises, it is fairly easy to perpetuate it.

On the other hand, the tomate's self-incompatibility makes it fairly likely that hybridization will take place with wild plants that may grow nearby. Hudson has found that there is no sharp demarcation between the wild and the domesticated plant. The variation in fruit size is more or less continuous, as one might expect if hybridization between the two types occurs frequently. Apparently other wild species will not hybridize with the tomate, so other sources of variability are not introduced.

The second part of the chapter title refers to the *pepino*,

Variation in fruits of pepino.

which is another domesticated plant of the Andes. Pepino is the Spanish word for cucumber, and some forms of the South American plant, hereafter simply called pepino, are green and somewhat elongated and do vaguely resemble the cucumber, which I suppose is the reason the Spanish called it by that name. Indian names, such as *cachun* in Peru, are also known for the plant, but today it is generally called pepino in most places, sometimes with a modifier to distinguish it from the true cucumber. The plant is fairly widely cultivated in the Andes at a variety of altitudes, and the fruits make their way to the markets. They are commonly eaten raw, and they have a rather pleasant sweet taste. Attempts have been made to introduce the pepino to other parts of the world, but without much success. I do not know if the lack of success results from difficulty in growing the plant or from failure of the fruit to become accepted. The pepino is now available, however, in several places in the United States through the efforts of a produce company, Freida of California, which imports it from New Zealand.

Whereas the cucumber belongs to the cucurbit family, the pepino, *Solanum muricatum,* is a member of the nightshade family. The plant itself somewhat resembles the Irish potato but does not produce tubers. The fruit shows a great deal of variation—in size, from 5 to 15 cm in length; in shape, from elongate-ovoid to nearly round; and in color, from green, yellow, or ivory to purple, and sometimes with darker-colored longitudinal stripes. The number of seeds in the fruits also varies greatly. The smallest fruits generally have a large number of seeds, the somewhat larger elongate ones usually have few seeds, and the large round ones may have no seeds at all. The last type of fruit appears to have the best flavor. Since it is seedless, propagation has to be vegetative, stem cuttings being used.

In addition to the great variation in the fruits there is also considerable variation in other characteristics. For example, the leaves may be simple or compound, and when they are compound, the number of leaflets may vary from three to seven. The stems and leaves may be nearly glabrous to quite hairy, and the flower color varies from white to deep blue.

The pepino has never been found growing outside of cultivation, and the wild plant that gave rise to it is unknown. I was interested in seeing if I could find any wild species that might be its ancestor, and I came up with two possibilities. One of these, *Solanum caripense* (named for the place, Caripe, in Venezuela where it was first collected by a botanist) is called *tzimbalo* in Ecuador and is fairly widespread in the Andes from Colombia to Peru at altitudes of 2,600 to 12,500 feet. This species has a number of characters in common with the pepino, but it also shows several differences, so I think it is best considered a distinct species. The berries are slightly smaller than a ping-pong ball. In my fieldwork I had difficulty in finding mature fruits, and subsequently I learned that was because the fruits are eaten, particularly by children. There is, however, not a whole lot to eat, for the berry is mostly juice and seeds. I also learned that the fruits are used in another, rather unusual, way. On visiting the section of the Quito market where medicinal plants were sold, I observed that the green berries had been strung on a thread to form a necklace. The vendor would not tell me how they were used, but from friends I found that they are worn by small children to protect them from *susto* or *espanto*. Both of these Spanish words are translated as "fright," but from the conversation I learned that they were worn for something more than that, for they served to ward off evil spirits. Sometime later I saw a small girl wearing a *tzimbalo* necklace, and although I would have liked to have taken her photograph, I did not try, for I realized that her mother, who was nearby, might associate me with the evil eye.

A second wild species rather similar to the pepino is *Solanum tabanoense* (from the place, Tabano, in Colombia). In contrast to the previous species, this one is somewhat rare, being found only in southern Colombia, southern Eduador, and one locality in Peru at altitudes of 9,200 to 11,500 feet. I thought it somewhat less like the pepino than *S. caripense* except for the fruit, which is very similar to that of the pepino. The fruit is larger (growing over two inches long) than that of *S. caripense*, has considerable flesh, and in taste more nearly resembles that of the pepino. Whether the fruits are eaten, however, I was not able to learn.

Anderson has found another wild species, *Solanum basendopogon* (the name referring to the hairs or "beard" at the base of the anthers), that shows some resemblance to the pepino and may have been involved in its origin. It is also rather rare, having been collected in only two departments in Peru at altitudes between eight thousand and ten thousand feet. Its fruit is smaller than that of *S. caripense*, and we also do not know if it is eaten by people.

A number of artificial hybrids have been made between the pepino and *Solanum caripense,* and some of these are fairly fertile. The only hybrid thus far secured between the pepino and *S. tabanoense* showed reduced fertility, but then the *S. tabanoense* parent also showed somewhat reduced fertility. Anderson secured a few seeds in crosses of the pepino with *S. basendopogon,* but they failed to germinate. It is still too early to say what the data on hybrids mean as far as the origin of the pepino is concerned, but they certainly do not eliminate the possibility that *S. caripense* is involved. Anderson has begun to look at certain chemical characteristics of these species, and the preliminary results indicate that both *S. tabanoense* and *S. basendopogon* are more like the pepino than is *S. caripense.*

There is no archaeological record of the pepino that might help establish its place and time of origin. The fruits, however, are represented on Peruvian pottery, so we know the pepino has a respectable age as a domesticated plant. That does not mean that it originated in Peru, of course. If the distributions of the wild species, however, are the same today as they were a few thousand years ago, it would narrow down the place of origin considerably if *Solanum tabanoense* or *S. basendopogon* were involved in the origin of the pepino. Of all the types of pepinos grown today, those from Colombia with small, many-seeded fruits are the nearest to what one might expect to find in a wild plant. On this basis one might conclude that the pepino originated in Colombia, but that does not necessarily follow, for we might also assume that a relatively unimproved form of the pepino spread widely after its domestication only to be later replaced at its center of origin by improved forms.

In view of the fact that there are three candidates for the progenitor of the pepino, no one of which satisfies all the requirements, could it be that it is of hybrid origin? Based on present distributions, there is no way in which natural hybrids could occur between *Solanum tabanoense* and *S. basendopogon,* but the distribution of *S. caripense* overlaps that of both of the other species, and, in fact, natural hybrids have been reported between it and *S. basendopogon.*

In the discussion of the tomate it was pointed out that that plant shows remarkably little variability. In the pepino, on the other hand, there is considerable diversity, and I think the difference between it and the tomate in this regard is easily explained. Although the first pepinos I grew were self-incompatible, Anderson has since found that self-compatibility is common in the species. This means that self-pollination is possible, and thereby variants can be rapidly stabilized and perpetuated. In addition, propagation of the pepino is often vegetative, which means that any mutant type can be readily perpetuated and clones established. There also may be another source of variation in the pepino that is not present in the tomate, for it is possible that the pepino hybridizes with several wild species. I think the introgression of genes from other species may account for the considerable variation in the leaves of the pepino. The diversity in the pepino is not a recent development, for we know from the descriptions left to us by the Spanish in Peru in the sixteenth century that the fruits were highly variable at that time, and we also know from their accounts that vegetative propagation was already being practiced.

CHAPTER 9

How Many Kinds of Peppers Are There?

To attempt to answer the question posed by the title of
this chapter, one first should define "kinds," which here,
of course, is used in the sense of categories. At the highest
level we are concerned with genera, and the principal genera that bear the common name "pepper" are *Piper* (family
Piperaceae), which includes black and white pepper and a
few lesser-known ones, and *Capsicum* (family Solanaceae),
which includes the red and green peppers, although other
colors are not uncommon. It is to the latter that this chapter will be devoted. Needless to say, when used without a
modifier, the word *pepper* can be a source of confusion.
Some years ago someone referred me to the editors of
Encyclopedia Britannica as an authority on peppers, and
they wrote me asking if I would contribute an article on
the subject for the encyclopedia. After an exchange of
letters, I learned that they wanted an article on *Piper* peppers, not *Capsicum* peppers. So I became an instant authority on *Piper* and a contributor to the encyclopedia. Another
example was provided by one of my students in my course
on summer flowering plants a number of years ago. I required the students to turn in a collection of plants properly identified with scientific names. This student included
a specimen of the garden sweet pepper, a *Capsicum*, but he
had labeled it *Piper nigrum*. I concluded that he had simply
looked up the scientific name of pepper in a book but had
not bothered to read the description given, and his grade
suffered accordingly.

When I began my study of *Capsicum* peppers, I don't
think I had ever eaten a really hot one. Soon I began to

taste them, partly through necessity, for I had to score the fruits of the plants that I was growing as pungent or non-pungent. Gradually I developed a liking for the mildly pungent ones, and today I would not think of eating bean or potato soup without adding a few drops of Tabasco. I cannot answer the question, What is the hottest pepper? for I have tasted many that are extremely pungent, and it would be impossible to say which one was the most pungent. My guess is that it would be some variety of *Capsicum chinense*. It is not my intention to go into the pungency in any detail here. I have treated the subject in my book on the nightshades, where I also discussed some of the unusual uses of peppers. Since that time I have learned of other uses, two of which deserve mention. It was my assistant, Lewis Johnson, who informed me that red pepper was used to flavor a Russian vodka. Naturally, in the interest of science I had to sample it. Recently from a former student, Cathy Greene, I learned that *Capsicum* powder was one of the ingredients in a foot warmer. I haven't yet tried it.

My interest in peppers began in 1946 when I was at the University of California at Davis and Paul G. Smith, a plant pathologist, brought a domesticated pepper to me for identification. Its identification was not a simple matter, but eventually I was able to put a name on it—*Capsicum pubescens*. By the time I had identified it, I felt that I knew the genus quite well from the literature, and I decided that it was a group that deserved more study. My decision was certainly influenced by the fact that the *Capsicum* peppers are native to the tropics and highlands of Latin America. I had wanted a good reason to get to the tropics ever since I had heard Bob Woodson, my professor at Washington University, lecture on the joys and trials of botanizing in the tropics. The group that I was then studying, sunflowers, was mostly confined to the temperate zone of North America, so I began a collaboration with Paul G. Smith on taxonomic studies of *Capsicum* that lasted several years. I made my first trip to tropical America in 1953 to

TABLE 1. Recent Classification of Domesticated *Capsicums*.

Smith and Heiser 1957	Hazenbush 1958	Heiser and Pickersgill 1969
Capsicum pubescens	*C. pubescens*	*C. pubescens*
Capsicum pendulum	*C. angulosum*	*C. baccatum* var. *pendulum*
Capsicum frutescens	*C. conicum*	*C. frutescens*
Capsicum chinense	—	*C. chinense*
Capsicum annuum	*C. annuum*	*C. annuum*

study peppers. Peppers have also provided research projects for several of my students, two of whom, W. Hardy Eshbaugh and Barbara Pickersgill, did their Ph.D. dissertations on *Capsicum*. Both have continued with their research on *Capsicum* to the present, and I have called upon their work in writing this chapter.

A brief survey of the history of the classification of *Capsicum* is necessary to understand the problems we encountered when we began our studies. In 1753 Linnaeus selected the name *Capsicum* for the genus and described two species, *Capsicum annuum* and *C. frutescens;* a few years later he added two more, *C. baccatum* and *C. grossum.* During the next century and a half more than eighty other species were named, some of which were wild species but most of which were domesticated plants. To many of the botanists of the time nearly every different type of fruit was thought to represent a distinct species. Then in 1898, H. C. Irish recognizing two domesticated species whereas others were following Bailey at the Missouri Botanical Garden made a detailed study of the genus and concluded that there were only two species, the original two proposed by Linnaeus. In 1923, L. H. Bailey of Cornell University, an outstanding authority on domesticated plants, went a step further and proposed that all of the domesticated peppers should be embraced within a single species, *C. frutescens.* When we began our studies we found that some botanists were following Irish and recognizing only one, which, needless to say, led to some confusion.

Our first clue that there might be more than one or two domesticated species of *Capsicum* came as the result of attempting to make hybrids between various accessions of peppers. We found that most crossed readily and produced fertile hybrids, but some were difficult to cross or would not cross at all, and some of the hybrids we secured showed reduced fertility. This examination of hybrids was followed by studies of the morphological characters of the plants, and using these in combination with the data from the hybridization study, we eventually concluded that five species of domesticated peppers should be recognized (Table 1). In addition to these domesticated species, four of which include wild or weedy forms, there are some twenty exclusively wild species in South America. Nothing more will be said about them, and the exact number of wild species and their taxonomy will have to await the conclusion of their study by Armando T. Hunziker, an Argentine botanist who for many years has been working with these species.

In a book of this sort it would be desirable to use common names for the various species, but they do not have recognized common names in English. In Mexico and much of Central America all *Capsicum* peppers are referred to as *chile*, whereas in much of western South America and the West Indies they are called *aji*. The exception is *Capsicum pubescens*, which is generally called *rocoto* in South America, but in Mexico this name is not used and it is called *chile mansana*. Thus I shall use Latin names.

1. *Capsicum pubescens*. This species is easily distinguished from all the others. It has black, wrinkled seeds, whereas the others have straw-colored, rather smooth seeds. The leaves are somewhat hairy and wrinkled or rugulose and have short stalks; the other species have nearly glabrous, smooth leaves and longer petioles. Its corollas are purple; the other species nearly always have light-colored corollas, white or greenish, although bluish corollas are also rarely found in *C. annuum*. *C. pubescens* is also ecologically distinct, growing at higher altitudes than the other species. It is cultivated only in the mountains of Latin America,

Fruits of *rocoto* (*Capsicum pubescens*) for sale at the market in Pisac, Peru.

particularly the Andes. A wild type of this species is not known, but it has close wild relatives in South America. It seems likely that it was domesticated somewhere in the Andes and from there was introduced into Central America. Its introduction into the latter area may not have occurred until after the Spanish had arrived in the Americas. Archaeological material of this species has been reported from highland Peru dated at around 6000 B.C., which would make it one of the oldest domesticated plants known for the Americas.

2. *Capsicum baccatum* var. *pendulum*. This pepper is readily separated from the others by the presence of brownish or yellowish spots on the base of the corolla that are lacking in the other species. The calyx also offers some distin-

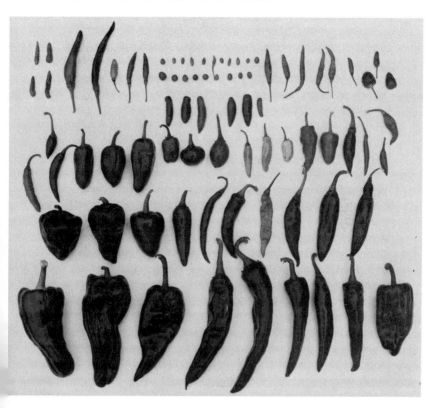

Variation in the chile pepper (*Capsicum annuum*). The small fruits without stalks in the upper center are from the wild variety. (From Heiser, *Seed to Civilization*, 2d ed., 1981, copyright © W. H. Freeman and Company; used with permission.)

guishing features, but if fruits alone are available, it is difficult to separate them from those of other species. This species is widely grown at altitudes lower than *C. pubescens* in South America, but somewhat surprisingly, it is little grown in other parts of the world. Its wild type *(Capsicum baccatum* var. *baccatum)*, found in Bolivia and surrounding areas, is a bird pepper, so called because birds eat the fruit. The archaeological record indicates that it was already cultivated by 2000 B.C.

3. *Capsicum frutescens.* This species, as a wild or spontaneous plant, is found throughout tropical America, extending
as far north as Florida. It is particularly widespread in
Brazil, where it is known as *malagueta.* The greatest variability in the domesticated types is found in Central America. The corolla in this species is usually greenish white,
in contrast to the white or purplish corollas of the other
species, and the fruits are often borne in pairs rather than
singly. One variety, Tabasco, is cultivated in the United
States, having originated at Avery Island, Louisiana, from
a Mexican wild pepper brought there after the Civil War.

4. *Capsicum chinense.* Why this pepper was named *chinense* has puzzled me, for it certainly is not from China.
It is the most common pepper cultivated in lowlands in
tropical South America, the West Indies, and much of
Central America. It is rather rare in Mexico, where it is
called *chile habanero,* probably indicative of an introduction
there from Cuba. Small-fruited spontaneous forms, which
are probably similar to the original ancestral type, are
found in northern South America. It is an extremely variable species, and some varieties are very difficult to distinguish from those of the preceding species. The one
character that generally separates *C. chinense* from *C. frutescens* is the presence of a ring at the base of the calyx, but
at times it is poorly defined.

5. *Capsicum annuum.* This pepper is the most widely cultivated in the world, and except for Tabasco, all the peppers
grown commercially in the United States belong to this
species, including the sweet bell peppers and pimientos
and several pungent varieties, including 'Chili,' 'Anaheim,'
'Cayenne,' and 'Jalapeño' as well as a number of varieties
grown as ornamentals. This species was originally brought
into domestication in Mexico. The wild type was another
bird pepper, *C. annuum* var. *glabriusculum,* which is found in
the southern United States, Mexico, Central America, and
the West Indies as well as northern South America. The extremely pungent fruits of the wild variety are still collected
and sold in markets in Mexico and Guatemala under the

name *chile pequín*. The wild type occurs very early in the archaeological record in Mexico, about 6000 B.C.

Smith and I published our principal papers on the species of *Capsicum* in 1953 and 1957. The question remained, of course, whether other botanists would accept our conclusion that there were five species of domesticated peppers or would continue to recognize only one or two. Confirmation that there were more than two species was shortly to appear. A Russian botanist, V. L. Hazenbush, quite independently of us was carrying out a taxonomic study at the same time as ours. His results were published in 1958, and he recognized four species (see Table 1). He failed to recognize *Capsicum chinense* and included what we would call by that name in part under his *C. annuum* and in part under *C. conicum*. His study did not include any genetic work, and his conclusions were reached solely on the basis of morphology.

Unfortunately, Hazenbush did not use the same names that we had used for two of the species. Obviously a species can have only one correct scientific name. So the next step was to determine which names should be used for those on which we disagreed. This work was done with Barbara Pickersgill, and although working with books and dried specimens was not as enjoyable as working with living plants, it was necessary. The rule of priority is used in selecting the correct name of a plant. Hazenbush used the name *Capsicum conicum* published by Meyer in 1818 (overlooking the earlier use of the same name by Lamarck in 1793) for the species we had called *C. frutescens*. The latter species, as we have seen, was published by Linnaeus in 1753, but we had to make sure that Linnaeus meant that name to apply to the same species we were calling by it. Linnaeus's description was very brief, as was customary at that time, and allowed for some uncertainty. To make sure of what he had in mind when he named it, we borrowed the actual specimen that Linnaeus had before him when he described the species, and after study of it we concluded that there was no doubt that the name *Capsicum frutescens*

belonged to that species.[1] Hardy Eshbaugh had already
made *C. pendulum* a variety of *C. baccatum,* another Lin-
naean species dating from 1767, which Hunziker had first
correctly interpreted, whereas *C. angulosum* of Miller, the
name used by Hazenbush, was not published until 1768.
Moreover, it is not clear that Miller's plant is the same as
what we were calling *C. pendulum.*

The question still remained whether four or five species
should be recognized. Several botanists in the Americas
have accepted our conclusion that there are five species,
but one European worker, A. Terpó, had followed Hazenbush's interpretation. Some work bearing on the species
problem has also been done since our studies, including a
study by M. J. McLeod with Hardy Eshbaugh and others.
They were able to measure the "genetic distances" among
the various species by a study of a number of enzymes by
means of electrophoresis. I shall not try to explain the
details, but the results showed not only that *Capsicum pubescens* and *C. baccatum* were very distinct from each other
as well as from the other species, but also that the three
other species were very similar, so similar that the re-
searchers could hardly justify calling them species. However, one should not make the decision whether they
should be called species on the basis of a study of enzyme
systems by themselves. Even before McLeod's study was
undertaken, however, I had concluded that it was difficult
to continue to recognize *C. frutescens* and *C. chinense* as
separate species.[2] In our original crosses between the two
species the resulting hybrids showed some reduction in
fertility, but since that time some fully fertile hybrids
between the two species have been obtained.[3] Furthermore,

[1] The fact that the specimens of the early taxonomists are often still
preserved is of inestimable importance to taxonomy. Dried specimens
of plants apparently will last forever if not destroyed by insects, fire, or
careless handling. Linnaeus's specimen of *Capsicum frutescens,* now on
deposit in Van Royen's Herbarium in Leiden, is called the type of the
species. Unfortunately, the use of the word *type* for such specimens has
led to some misconceptions. That it is so called does not mean it is

and perhaps more decisive, the morphological differences between *C. frutescens* and *C. chinense* are, at the best, very slight, and I am not sure they are consistent. While I can usually refer living plants to one or the other species, it is often impossible to do so with dried material. From both practical and biological standpoints I can see no justification for recognizing them as separate species.

If the reader has found parts (I hope not all) of the foregoing rather boring, I understand, for I did not find all of it terribly interesting to write. Such taxonomic work is nonetheless important, not only to taxonomists but to others as well. For example, as a result of a better understanding of the species, plant breeders can now plan their experiments with greater exactness. However, I think certain other aspects of our work are of greater significance than the taxonomic conclusions. As a result of our studies we were able to identify the wild ancestral types of several of the species, and it is thus apparent that three species had their origin in South America and that one species had its origin in Mexico. This should be of interest to anthropologists, for it demonstrates that there were separate origins of domesticated peppers and not a single place of origin in Mexico, as some had previously supposed. This knowledge, combined with that from other domesticated plants, indicates that there was probably more than one origin of agriculture in the Americas.

The presence of wild types in the various species of pepper also allows us to visualize the changes that have

---

typical of the species, but only that it serves for the attachment of the name.

[2] A colleague of mine, Frank Young, when he heard that I was recognizing four species, pointed out that after many years of study I now had the taxonomy the way Linnaeus had it two hundred years ago. Not quite true, but not far from the mark.

[3] As the reader may already have gathered, data from hybrids have to be interpreted with caution in relating them to taxonomy. It has also been found that some closely similar forms of *Capsicum annuum* when hybridized produce progeny showing some reduction in fertility.

occurred with domestication. The wild species all have very small red fruits which are borne erect and which are deciduous or easily removed at maturity. All of these characteristics are probably adaptations for dispersal by birds. The erect red fruits attract the birds to the plant, and the small size and ready removal of the fruits make them easy for birds to gather. The wild peppers also all have extremely pungent fruits, and whether this is attractive to birds is not known; all that can be said is that it apparently does not hinder their harvest by some birds. With domestication we find larger sizes and variable shapes of the fruits, several colors in addition to red, fruits often pendent so that they hang among the leaves, nonpungent as well as pungent types, and persistent instead of deciduous fruits. Several of these characteristics make the fruits less subject to predation by birds and hence are important to people who would like to have all of the fruits for themselves. With the exception of an overall increase in size of various parts of the plant, all of the characters that distinguish the wild from the domesticated peppers reside in the fruit. That there is great variability in the fruits of domesticated peppers is not unexpected, for Darwin in the last century pointed out that the greatest variation in a domesticated plant generally occurs in that character for which humans cultivate it. Human beings have promoted the great variability in the fruits apparently by conscious and unconscious selection.

Enough digression — it is time to return to our question of how many kinds of peppers there are. We have seen that there are four (or five?) species, and each species contains a great array of fruit types, differing in shape, size, color, position, and pungency. I think it is to these that people refer when they ask how many kinds of peppers there are.[4] In the last century many of these were named as species,

---

[4]These are not straw people, for I have actually been asked this question — not often, true, and not by many botanists, who perhaps realize the impossibility of an answer.

Pepper row in the market at Chichicastenango, Guatemala.

and more recently some of them have been dignified with Latin names as varieties. To distinguish naturally occurring varieties from those of horticultural origin, it is now recommended that the latter be designated as cultivars (= cultivated varieties) and given a fancy rather than a Latin name. Thus we have, for example, *Capsicum annuum* cv. 'California Wonder' and *C. frutescens* cv. 'Tabasco.'

It is impossible to say exactly how many cultivated varieties there are. First of all, not all the varieties have been cataloged and named. Many grown by Indian tribes in South America fall into this category, although the Indians will usually have their own names for them. In fact, for the most part only those varieties grown in Europe, the United States, and Mexico have been scientifically named. Second, there is some disagreement among botanists and horticulturists about what should be named. Even if there were agreement about the varieties to be named, the resulting list would soon be out of date, for in the United States and several other countries almost every year new varieties are produced by plant breeders through intentional hybridization and selection. Some of these cultivars, however, may differ only very slightly from previous ones, for example, by being resistant to some diseases or by earlier maturation. In many tropical countries new varieties may come into existence through spontaneous hybridization between existing varieties or through gene mutation. Some of the older cultivars may disappear, but many of them come down to us little changed after hundreds of years. Some of the Mexican varieties, for example, appear to be the same as those described by the Spanish shortly after they entered the country in the sixteenth century.

An examination of some of the classifications of the varieties of *Capsicum annuum* will give a better appreciation of the situation. The first extensive treatment was that of Irish, who in 1898 named 43 different varieties. In 1932, A. T. Erwin of Iowa State University in a classification of the varieties in the United States recognized 153 different ones. The increase in number was largely because of an increase in the number grown in the United States, some developed since the time of Irish's study, and some introduced from outside the country, but in part from a different taxonomic interpretation. In the most recent classification known to me, Terpó in 1966 gave formal recognition to only 33 types. There is, therefore, hardly agreement about how many kinds of peppers there are in *C. annuum.*

There are no similar classifications for the other species. From my own experience in Latin America and in growing peppers I would judge that the *chinense* types of *Capsicum frutescens* probably are as variable as *C. annuum,* and there is also considerable varietal diversity in the other species. There can be no definite answer to the question that we started out with, except to say that there are lots of different kinds of peppers.

CHAPTER 10

Peperomias

In my travels to Latin America I have seen a large number of beautiful plants. Some of them have already been grown as ornamentals in various parts of the world, but there are others that are presently unknown in cultivation. I have always felt that, along with trying to improve some of our food plants, I would like to introduce a new ornamental to add a touch of beauty to people's lives. On my trips I have always kept my eyes open for worthwhile subjects, and over the years I have collected seeds of a number of plants that appeared to have something to offer either in their flowers or vegetative appearance. Most of my collecting has been done in the highlands, and unfortunately, not one of the plants that I have brought back from there has has yet shown much promise when I have grown it in Bloomington, Indiana, either in the greenhouse or the garden.

There are two primary reasons for my failures. First of all, the plants that come from the highlands grow where it is always relatively cool, and they frequently will not tolerate the hot summers of Indiana. Another reason is that many of my collections come from near the equator, where the days and nights are approximately the same length. It is well known, of course, that day length is an important factor in the flowering of many plants. Of course, twice a year the days and nights are the same length here, but that may not be enough to induce flowering. Even when my highland species do flower, they may fail to put on much of a show, and often they fail to produce fruits— sterility being promoted by the high temperatures or the day length or a combination of the two.

155

My luck has been a little better with the lowland tropical plants. In 1967, when I was in Costa Rica, I visited
a cacao (chocolate bean) plantation with Jorge Soria, and I
found that the trunks of the trees were covered with a creeping small-leaved *Peperomia* which I found most attractive.[1]

The *Peperomia*s, which belong to the same family as
black pepper *(Piper nigrum)*, are grown as house plants
for their foliage. The leaves are somewhat succulent, and
why this should be I am not sure, for most succulents come
from deserts or other dry areas. A few *Peperomia*s are
desert plants in Peru, but most species grow in extremely
wet areas. Their succulent leaves may be an adaptation to
water stress, which may result at times because they have
poorly developed root systems or grow upon other plants
as epiphytes. The extremely small flowers are borne in
short, slender spikes, which add to the attractiveness of
some of the species. The most appealing thing about the
Costa Rican plant was the numerous nearly round leaves,
which were only one-third inch in diameter somewhat like
those of baby's-tears (*Helxine soleirolii*) of the nettle family.
Moreover, its creeping nature would qualify it as a good
plant to grow in hanging baskets with the leaves trailing
over. I also found that the leaves had a pleasant spicy odor
when they were crushed. Many species of *Peperomia*, I later
learned, have this characteristic. In fact, one of them, *con-
gona* or *congonilla* (*Peperomia inaequalifolia*), is used to flavor
drinks in Ecuador.

After my return to the United States a couple of days
later, I had no difficulty in getting the plant to grow in
pots, and people who saw it found it attractive. I thought
I had a most worthwhile introduction. My judgment proved
to be correct, but my timing was not, as I found when I

[1] Although Bailey gives the name "radiator plant," the *Peperomia*s do
not seem to have an accepted common name, which is perhaps just as
well, for I feel that we should use the genus name rather than an artificially contrived common name. Some people claim that genus names
are "difficult," whatever that means. It certainly has not hindered the
acceptance of some of them, such as *Petunia.*

A collection of cultivated *Peperomias*. *Peperomia caperata*, one of the most popular species, is in the center.

began to identify it. It keyed readily to *Peperomia rotundifolia*, and I learned that it had been introduced into horticulture a hundred years earlier.

In 1982, when I was on a visit to the hacienda of Luis Morales near Tena in the eastern lowlands of Ecuador, I found another plant that appeared to have promise. This one was growing on a nearly rotten fence post, and unlike the previous one, which grew in great abundance, I found only a single plant. I removed a few spikes, and although I could not see any seeds in them—the seeds of *Peperomias* are quite small—I hoped for the best.

After I returned to Bloomington a month and a half later, I planted the whole spikes. Nothing happened for a long time. Finally, after three months, I saw that three seedlings were emerging. Two of them proved to be something other than *Peperomia*. I still have them growing, for they are also rather ornamental. They have not yet flowered, so I have no idea about their identity, but they are certainly not any of the usual contaminants that sometimes come up in pots in greenhouses. How the seeds could have gotten into the envelope with the *Peperomia* spikes also remains a mystery. The third plant was the *Peperomia* I had hoped for. It grows well, and starts are readily made vegetatively. The plant proved to be as attractive in a pot as it had been in nature.

After the plant produced spikes, I set out to try to identify it. The identification of a plant from Ecuador is often no simple task, for there is no flora of the country with descriptions of the species and keys for their identification (a deficiency that is shared by other countries in Latin America).[2] A flora of Ecuador is now being prepared by Gunnar Harling and Benkt Sparre of Sweden, but it will be many years before it is completed. Fortunately, however, the Piper family in northern South America had been the subject of a detailed study by William Trelease of the University of Illinois and Truman G. Yuncker of DePauw University. The study was made from a large number of dried specimens, borrowed from herbaria in both Europe and the United States, including the original material, or type specimens, on which each species was based. Keys to identify the species are given, along with complete descriptions, the known geographical range, and photographs of most of the species. This was a tremendous undertaking, involving many years of work.

Three hundred and sixty-five species of *Peperomia* are included. Many of the species are described for the first

---

[2]There is also no flora of the United States, but there are many regional or state floras so that plants from most parts of the country can be readily identified. The lack of even an inventory of the plants of many regions is most unfortunate. With the rapid destruction of the vegetation now occurring in the tropics, many species are becoming extinct even before they are known to science.

Peperomia rotundifolia in the greenhouse. *Peperomia polybotrya* may be seen on the right.

The "new" *Peperomia* from Tena in cultivation.

time by Trelease and Yuncker, and many of the others
are known but from a single locality. It is not unusual
to find many new species when a tropical group comes
under careful study, nor is it too surprising to find a great
number known only from one place, reflecting the lack of
extensive collections in the tropics. On the other hand,
I could not help but wonder whether Trelease and Yuncker
had a very narrow species concept and that other taxonomists might go over the same material and conclude that
there were far fewer species in the genus. Dr. Yuncker in
the introduction to the monograph admits, "The number
of new species recognized is extraordinarily large and
perhaps appears unreasonable. . . . It is believed likely that
careful field study on more abundantly intergrading materials will reduce" the number of species. At the same time,
he predicts it probable that more undescribed species will
be discovered.

Unfortunately, we as yet know little about the biology
of the species. Trelease and Yuncker were concerned primarily with their names and their morphological descriptions. Information about pollination and seed dispersal is
almost nonexistent. We know very little about the ecology
of the species. Do they coexist in nature, and does hybridization occur between them? Such information could
help us understand the apparently large number of species in the genus. For example, if hybridization does occur,
perhaps some of the specimens described as species are
nothing more than hybrids. Such studies would involve
detailed observations in the tropics, supplemented with
greenhouse or laboratory studies. Some might criticize
the work of Trelease and Yuncker for not giving us such
information, but their type of taxonomic work was necessary to set the stage for future investigators. In fact, we
need much more such "old-fashioned" taxonomic studies
for many groups. Unfortunately, few people carry out such
studies today, no doubt because most investigators find
the study of the biology of species more interesting than
their classification, and I, of course, in view of my own

interests, am hardly in a position to criticize them. Moreover, taxonomists in universities, as compared to those in
herbaria and museums, are seldom rewarded by promo-
tions or higher salaries for carrying out such studies.

But I must return to the *Peperomia* from Tena. After
going through the work of Trelease and Yuncker, I concluded that it had to be either *Peperomia urocarpa* or *P.
serpens* and that it was more likely the latter, which had
previously been collected from Puyo, which is very near
Tena. Then, to ascertain whether the species was in cultivation, I consulted *Hortus* III and *Exotica*, and I found that
both of these species are given, so once again I was not the
first to recognize the merits of this plant. However, the
plant from Tena does not completely agree with the description of the cultivated plants given in *Exotica*, so perhaps my collection represents a different race and still
deserves cultivation.

The *Peperomia* story does not end here, for I have not
yet given up on finding one new to cultivation. When I
was in Ecuador in 1983, I made a visit to the Hacienda
La Suiza. The hacienda, which is owned by Sergio Soria,
is near Patate at an altitude of over ten thousand feet.
Although most *Peperomia*s are found at lower elevations,
quite a number also occur in the highlands. As I was hik-
ing in the undisturbed forest at La Suiza, my search was
rewarded with the discovery of a most striking plant. It
was quite small and had dark red spikes and stems and a
reddish leaf undersurface. It was, I decided, quite attractive and worthy of introduction to horticulture.

When I returned to Bloomington, I could find nothing
like it among the cultivated species listed in *Hortus* III or
Exotica, nor did I find any description that agreed with
it in the work of Trelease and Yuncker. This, of course,
might mean that it was an undescribed species, new to
science. Before concluding such, however, it was necessary
to determine if it had been described since 1950, the date
at which their work had been published. So I went to the
Kew Index, which gives all the new flowering plant spe-

cies published, and I found that over 150 species had been described since 1950, mostly from tropical America, many of them from Ecuador. Dr. Yuncker had continued to describe new species until the time of his death in 1962. If he were still living, I would send the specimen to him for identification. Unfortunately, I know of no new authority on the genus, which means that I shall have to find a name for it myself. This will involve going through the descriptions of the new species published since 1950, which are scattered in many journals, and then it may also be necessary to borrow herbarium specimens to make detailed comparisons. I would not be surprised if it turned out to be a new species, for to my knowledge no botanist has ever collected in that area of Ecuador before.

As I write, it has been over three months since I planted seeds of the species, and they have not yet germinated. I am afraid they were immature and that I shall have to make another trip to Ecuador, a prospect that is not at all displeasing to me. Of course this *Peperomia* may not prove to do any better than the other highland species that I have previously tried to introduce, but I feel that it is worth another effort.

CHAPTER 11
Sumpweed

One would hardly expect any plant with a name like sumpweed to be beautiful. In fact, it might be described as rather unattractive, but I find it a most interesting plant for reasons that I shall develop shortly. Sometimes the name "marsh elder" is used for it, and I think that gives a little more dignity to the plant. Both of these names I have taken from books, for the few people whom I have met where I have collected the plant had no name for it at all, and I find that few botanists have much of an acquaintance with it, even under its scientific name, *Iva annua*. It is, however, well known to American archaeobotanists, for the plant figures rather prominently among the prehistoric food plants of eastern North America. Moreover, it is one of the few plants—perhaps the only one—that was domesticated but became extinct as such before historical observations could be made on it.

The plant is a member of the composite family, or Asteraceae, which includes such familiar plants as sunflowers, daisies, dandelion, and ragweed. Of these, sumpweed is most nearly like ragweed, as seen in the similar flower clusters, but whereas our common ragweeds have deeply divided leaves, those of sumpweed are merely coarsely toothed. Like ragweed, it is wind-pollinated, and the pollen is known to cause hay fever where the plants are found in some abundance. Nearly all parts of the plant have a strong camphorlike odor that some people find objectionable. In fact some people who handle the plant claim that it causes a skin irritation, although it has never so affected me. The main subject of this chapter, however, is the fruit and seed. The fruit is an achene and encloses a single seed.

The plants are annual herbs, usually about two or three feet tall where I have seen them, but they are reported to reach heights of seven feet. As the colloquial names imply, the plants are generally found in wet places — floodplains, margins of ditches, and the like. The plant is local throughout many places in the central United States, with the greatest concentration from Illinois to Nebraska, south to Alabama and northern Texas.

I have never made a serious study of the plant, but I have collected it in nature and have grown it in the greenhouse and garden off and on for more than thirty years. I had mentioned the genus to several students as a possible dissertation topic, and ultimately a former student, Raymond Jackson, wrote a monograph on it in 1960. In addition to bringing the biological information concerning the plant together, he was responsible for restoring the original scientific name to sumpweed. Although the plant was named *Iva annua* by Linnaeus in 1753, another botanist, Willdenow, apparently unaware that Linnaeus had named the plant, designated it as *Iva ciliata* in 1804, and the latter name was widely used for the plant until the time of Jackson's study. Much of the archaeological study of the plant has been done by Richard Yarnell and Nancy and David Asch. The Asches carried out archaeological investigations of it in Illinois, made detailed studies of the plants, and brought out a comprehensive review in 1978. I have depended heavily on these people for the account here.

The story begins several thousand years ago in the central part of what is now the United States. The people at that time depended entirely on wild sources of food. Deer, small mammals, waterfowl, and fish were the animal sources, and various herbs — also were important. Among the latter were lamb's-quarters (*Chenopodium bushianum*), smartweed (*Polygonum erectum*), May-grass (*Phalaris caroliniana*), sunflower, and sumpweed. Sumpweed appears in the archaeological record at about 3500 B.C. in Illinois. The achenes are about 3 mm long and 2 mm wide, which is clearly

Sumpweed (*Iva annua*) in southern Indiana.

within the size range of the modern wild sumpweed, so it is most likely that the collections came from wild plants.

About 2000 B.C. larger achenes appear in archaeological deposits, and an increase in size continues up to A.D. 1300. Achenes now average 5 mm by 3 mm from many sites, and from one population we find achenes 7.6 mm by 5.2 mm. These larger achenes are clearly outside the range known for wild plants, and although we have nothing more to go on than the achenes, the conclusion seems inescapable that human beings had something to do with the increase in size. Experiments have shown that if wild plants today are grown under ideal conditions—plenty of light, water, and fertilizer—there may be only a slight increase in achene size. Thus the larger achenes from the archaeological record must be a result of genetic differences. These large achenes are fairly widespread. They are most abundant in Kentucky, Missouri, Arkansas, and Illinois and are also reported from Iowa, Tennessee, North Carolina, and Mississippi.

Although at one time it was speculated that the achenes of sumpweed may have been used for some purpose other than as food—for medicine, for example—it is now generally thought that they were consumed as food. The finding of achenes in human feces at one site shows that, at least on one occasion, they were eaten. The Asches carried out experiments to determine how primitive man may have collected and used them for food. First they found that the plants could be readily harvested by stripping them by hand. They estimate that under certain conditions a person could obtain enough achenes in one hour to satisfy an adult's daily energy requirement.

The seed, as it comes from the plant, is covered by the tough, fibrous shell of the achene, and the Asches postulate that the shell was removed before the seeds were consumed. How could prehistoric man have done so? Pounding the achenes was probably not the answer, for the oily seeds would probably be crushed as a result, making it difficult to separate them from the shells. The Asches report that boiling the achenes causes the shells to weaken or split. Then, after they have dried, the achenes can be rubbed between the hands to separate the shells from the seeds. Winnowing then separates the shells from the seeds. They found the boiled or roasted seeds have a pleasant nutlike taste, and the objectionable odor is eliminated.

How the prehistoric people processed the plant is probably more easily explained than why they domesticated the plant in the first place. If the plants grew in some

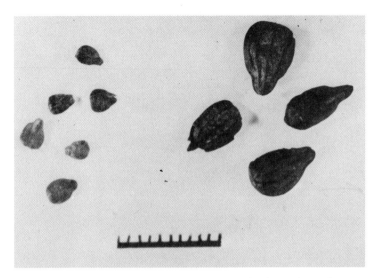

Achenes of present-day wild sumpweed (*left*) and achenes from the Turner site in southeastern Missouri, dated at about A.D. 1300 (*right*). (Archaeological material, courtesy Paul Minnis.)

abundance in nearby floodplains, why did the people ever bother to plant them? It could be, of course, that they were not abundant or that people who lived some distance from them wanted them at hand. The same problem awaits explanation for many other plants, and I shall address it at greater length in the last chapter.

One thing we do know is that the people would not necessarily have had to plant sumpweed in a floodplain or lowland site. Although in nature it is nearly always found in wet places, it will grow successfully in upland sites, as both the Asches and I have observed. Moreover, it will reseed itself in such sites. Perhaps achenes dropped around campsites thousands of years ago gave rise to plants, and this inspired man to drop seeds intentionally for the first cultivation of the plants. Then with continued cultivation the plants became domesticated, with an increase in the size of the seed.

The Asches have postulated that the cultivation may

have been in upland sites. They suggest that such would have been necessary in order to inhibit gene flow between the naturally occurring plants and the cultivated ones so that the domesticated form could arise and become fixed. Probably such isolation would allow a more rapid evolution of the domesticate. Yet, even if the wild and cultivated plants were in contact, disruptive selection could occur, if the people were deliberately selecting larger achenes for planting, to allow the origin of the domesticate. Such disruptive selection is known to occur between wild and domesticated *Sorghum* in Africa even though it, like sumpweed, is wind-pollinated.

It seems reasonable to suppose that people were practicing some form of selection in order for the domesticate to arise. The easiest way to account for it might be to assume that they were selecting larger achenes for planting. For them to have done so implies some knowledge of genetics several thousands of years before the science was born. Perhaps they realized that larger people tended to have larger children. Selecting for larger achenes, however, would probably not have any, or much effect, on the next generation. It would probably take selection over many generations before the results were apparent, for larger achenes are probably controlled by several genes. Why would people continue to select larger achenes if they didn't have immediate rewards? We cannot, of course, be sure that they were selecting larger achenes, and it may be that some other aspect of the plant, such as greater plant size, was the object of their attention. If larger plant size were correlated with larger achene size, which does not seem too unlikely, then we could also account for the eventual development of the larger achenes. Unfortunately, we have no idea what the domesticated plant looked like. Only the achenes have been left for us. One wonders in what other ways if any, the domesticated plants differed from the wild ones. Did they tend to retain their fruits at maturity as many other domesticates do, or did they fall from the plant fairly rapidly as on the wild plants?

The former, of course, would have been an advantage to the people, for they would get more achenes when they harvested the plants.

Could larger achenes have resulted without direct human selection? Let us assume that the people did grow the plants in an upland site, that mutations for larger achenes occurred or were already present, and that the larger achenes conferred an advantage to the plants in the new habitat but not in the original wet habitat. The advantage could be that the larger achenes produced more vigorous plants that could have competed better with the other vegetation that might occur in the new habitat. The more vigorous seedlings of sumpweed would outgrow the less vigorous seedlings from the smaller achenes, if the latter survived at all. Thus there would be more plants from the larger achenes available for the harvest in the fall. In turn, there would be more larger achenes available for next year's planting without any selection on the people's part, somewhat like the case postulated for the tomato (chapter 8). Other models might be set up to explain the larger achenes, but they would involve a number of assumptions, just as does the one proposed here. Experiments might be set up to test this hypothesis, but for the present I still feel that it would be difficult for the achenes to increase to the size they did without conscious human selection occurring at some time. However, in the domestication of many plants it is postulated that some form of unconscious selection by human beings may have been involved.

Both Yarnell and the Asches have suggested that the domesticated sumpweed may have been a tetraploid—that is, a plant with twice the number of chromosomes of the wild type. Their suggestion comes from my work. Polyploids occur in nature from various causes, and they can be produced in the laboratory fairly readily by treating plants with the chemical colchicine. Some years ago when I was treating germinating seeds of sunflowers with colchicine, I had some sumpweed germinating at the same time, so I treated them also. As I expected, the polyploid plants that resulted produced achenes considerably larger, about 4.4 mm by 3.0 mm, than those of the parent plants. The plants, also not unexpectedly, were nearly sterile, producing only a very few achenes. Such reduction in seed set is not uncommon when the chromosome number of individual plants of a species is doubled. The fertility of the polyploid plant might be improved by selecting achenes from the most fertile plants over a number of years, or a polyploid produced from hybrids of different races of the species might be expected to show higher fertility. The latter could have occurred, for example, if people had brought achenes of the plant from some distant area to a new area where the species already grew. Subsequently, a hybrid between the two races might occur, and if this plant doubled its chromosome number, the people might realize the superior value of its larger achenes and save them for planting.

While it is true that many of our domesticated plants— wheat, oats, sweet potatoes, and Irish potatoes among them—are polyploids, many are not. Corn, beans, squash, and sunflower, all diploids, have seeds or achenes that are considerably larger than those of the wild types thought to have given origin to them, so it is clear that the increase in size of the seeds of these plants was done without the involvement of polyploidy. Although it is possible that the domesticated sumpweed was a polyploid, I don't think it is any more likely than an increase in size of the achene solely through gene mutation and selection.[1]

At the time people were domesticating the sumpweed, squash was already a domesticated plant in central North America. The sunflower was also being domesticated at

[1]After the above was written, I had ten plants in the garden produce a few branches about twice the ordinary size along with normal branches. The abnormal branches had achenes considerably larger than the normal branches. Cytological examination revealed that the large branches were diploid, so the cause of the large size is unknown. Nor is it yet known whether the seeds from the gigantic branches will pass this trait on to their offspring.

about the same time as sumpweed. We don't know if the same groups of people were domesticating both plants, but it is a possibility. Corn had probably not, and beans certainly had not, arrived from Mexico before the domestication of the sumpweed and sunflower began. Could the arrival of corn and beans from Mexico have something to do with the demise of sumpweed as a domesticated plant?

As we have seen, sumpweed as a domesticated plant had a fairly wide distribution, but it was nowhere observed, or at least recorded, by the Europeans who arrived in the eastern half of North America. Some of the archaeological finds of it are rather late: A.D. 1250 to 1450. It has been speculated that it may still have grown in some places when the Spanish arrived, and had some of the early explorers taken different routes, they might have seen it. The early visitors found the agricultural Indians growing corn, beans, squash, and occasionally sunflowers. Perhaps with these plants sumpweed was no longer needed, for together they gave a well-balanced diet. Sumpweed provides an oily seed, but so does the sunflower, which may have been easier to grow and certainly was easier to harvest. Moreover, corn, beans, and squash could be grown together, and sumpweed would probably have had to be grown as a monoculture. The sunflower managed to hold on as a minor crop because of its oily seed and for other uses, such as parts of the plants for pigments. It seems probable that the greater work involved in growing, harvesting, and processing sumpweed compared to the other crops led to its decline. There is still another possibility that should be explored, however. While we may like to think of the "noble savages" as not being prey to all of the ills that plague us today, probably they, too, suffered from pollen allergies. When sumpweed was cultivated on a large scale, perhaps some people connected the sumpweed with their runny noses and itchy eyes. The sunflower, not being wind-pollinated, did not cause allergies, so it was kept and sumpweed was abandoned when other plants arrived.

We can be fairly certain that sumpweed did not disap-

pear because it had little food value. The Asches had
seeds analyzed, and the results indicate that it is indeed
a nutritious food. Calorie content is similar to that of
sunflower seeds, less than that of most nuts but superior
to that of starchy seed and root crops. The seeds are very
high both in fat, 40 percent, and protein, 32 percent. True,
the protein is low in the essential amino acid lysine, but
then so are all of the cereals and many other plant foods.
Probably few other foods equal sumpweed in vitamins
and minerals; it is a particularly good source of thiamine,
niacin, calcium, iron, and phosphorus.

In a world desperately in need of more and better food,
one wonders if experiments should not be undertaken to
attempt to restore sumpweed as a crop. Sometimes students
ask me why we do not attempt to domesticate new sources
of food today. Little is being attempted along this line,
for it is realized that it would take tremendous investments
to bring a wild plant into cultivation and bring its yield
up to that of our standard crops. With sumpweed, however,
we know that we could make it a domesticated plant. At
least prehistoric man did so. Methods could probably be
devised fairly readily to harvest and process it efficiently.
Its high oil content suggests that it could be made an important source of vegetable oil. The Asches' analysis did
not reveal the nature of the oil, but other oil plants from
the same plant family—safflower and sunflower—have un-
saturated oils highly desirable for food use. Probably the
greatest need of better vegetable oils today is in the tropics,
and unfortunately the sumpweed does not appear suitable
for those regions. As is true of many temperate-zone species, long nights are necessary to induce it to flower.
Another possible disadvantage of it as a crop is that large-
scale cultivation of it might cause allergies, so it might be
necessary to grow it in areas somewhat removed from
people.

Chapter 12

A Plague of Locusts

The honey locust, *Gleditsia triacanthos,* is probably well known to many of my readers, if not in nature at least in cultivation, for it is easily grown and widely planted as a shade tree or ornamental. Those who have not seen it in the wild in the eastern United States, however, have missed one of its most interesting features, its "ferocious" thorns. Such armor is hardly appreciated in the yard, and several cultivated varieties have been developed that lack spines. Also, those who know only the cultivated plant may have missed another of its prominent features, the rather long and usually somewhat twisted pods. Although Sargent in 1890 wrote that the pistillate ("female") trees are more desirable as ornamentals than the staminate ("male") for "the fruit which hangs in great profusion from all the branches are conspicuous and beautiful from midsummer until they fall," today it is the staminate forms that are much more widely grown. The reason for this, I suppose, is that when the pods fall they make a messy lawn.

The leaves of the honey locust come out rather late in the spring and, being rather lacy, do not give as much shade as do those of many other trees, but this is an advantage in that a lawn often does well under them. The wood of the tree is hard, strong, and durable and has a number of uses. I find that it makes good firewood. The pods are filled with a sweetish pulp and are much sought after by many animals, and a beer is said to be made from them. In 1928 the American Genetic Association had a contest to find the "best" honey locust, and the winner came from Georgia; its pods had a sugar content of nearly 30 percent. In his book *Tree Crops,* J. Russell Smith in 1950

Honey locust in fruit.

wrote enthusiastically about the honey locust and advocated its use as a crop plant. Since that time I have seen articles suggesting that greater use of the pods be made for livestock feed, and even for human food, but so far I think little has been done about it.

A second species of the genus, the water or swamp locust, *Gleditsia aquatica,* is also found in the eastern United States, and as the name implies, it grows in wet places. It differs from the honey locust primarily in the pods, which are one-seeded and dry instead of being filled with pulp. This species is also characterized by stout thorns, and the fact that unarmed forms are not known probably explains why it is not cultivated. Except for the thorns, I think it is as desirable as the honey locust. It is sometimes cultivated in botanical gardens.

A third species, *Gleditsia texana,* based on material from

Spines of honey locust.

Texas, was once recognized in the United States. It is still found, a rather rare plant, but now it is generally thought to be a hybrid of the previous two species instead of a species in its own right.

The genus *Gleditsia* is known from the fossil record in the United States. In addition to the species of North America, it also has representatives in Asia and South America, which probably indicates that the genus is of considerable antiquity. These other species, some occasionally grown as ornamentals, need not concern us further here.

The scientific name of the honey locust, *Gleditsia*, was chosen by Linnaeus to honor Johann Gottlieb Gleditsch, a contemporary botanist; the specific epithet, *triacanthos*, means "three-thorned." As is often the case with common names, however, that of the honey locust is not so easily explained. The word *locust* apparently was first used for certain insects similar to our grasshoppers. Then because

Water locust with fruit.

of the resemblance of the pods to the insect, it was used for the fruit of the carob tree, an Old World tree which belongs to the legume family, to which the honey locust also belongs. This is not my explanation but comes from the Oxford English Dictionary. Some have thought the carob pods were the locusts of John the Baptist. Later, when the English found a tree in Virginia with a somewhat similar pod, they called it "locust." The honey part, I would guess, comes from the sweet pulp of the pod.

My introduction to the honey locust problem[1] came in the fall of 1955 or 1956 when a man entered my office and, without introducing himself, asked — or perhaps "demanded" is a better word: "When are you botanists going

[1] Please note that I have refrained from calling it a thorny problem.

to do something about that honey locust?" I learned later that the man was Scott McCoy, a high school science teacher from Indianapolis.

He seemed to think that I should know all about "that honey locust," which I didn't. Piecing the story together from what he told me that morning and from a study of the specimens and letters in our herbarium, I learned that in 1937 McCoy had collected an unusual specimen of honey locust near Emison, Indiana. He sent his specimen to the Indiana botanist Charles Deam, who in turn sent it to E. J. Palmer, an authority on trees, at the Arnold Arboretum. Palmer wrote Deam, "I think someone would be justified in describing it as a new species if he had flowers and complete material for a description." McCoy later returned to the spot where he had collected the specimen only to find that the tree had been cut down, and he could find no others like it in the vicinity.

I don't recall exactly what I said to McCoy that morning, but I think it was words to the effect that it was probably just an extreme variant of the honey locust or possibly a hybrid derivative of that species with the water locust.

With that I forgot about honey locusts until 1958, when McCoy described his specimen as a species new to science in the *Proceedings of the Indiana Academy of Science.* I was a little upset to see it published, for I hardly thought he had carried out sufficient research to reach any conclusion. Or perhaps, as some have suggested, I was annoyed that an amateur, not a professional taxonomist, had done it. I hasten to add that I welcome the work of amateurs to botany, and certainly in the past they have made many valuable contributions. In fact, Charles Deam, mentioned earlier, who did a treatment of all the higher plants of Indiana, was an amateur, and his *Flora of Indiana* is a superb book. His work was so thorough that only a few species escaped his attention, and it is unlikely that a honey locust would have.

McCoy's new species differed from *Gleditsia triacanthos* only in the shorter, straighter pod which lacked pulp.

Surely this was hardly the basis for a new species, but maybe I was wrong. So I resolved that I would have to learn more about honey locust, and that was a wonderful excuse for some field trips to southern Indiana. My research work at the time was largely limited to the greenhouse and laboratory, and one of the reasons that I had gone into botany was the appeal of fieldwork.

My first trip was with Don Burton to learn more about the water locust, for I had never seen it outside of the herbarium. We stopped along the way to look at honey locust, so it was rather late when we got to the Little Cypress Swamp where the water locust grew. In fact, it was so dark we could hardly distinguish one kind of tree from another, and we didn't have a flashlight. So how does one find a locust tree in the dark? Obviously one goes around feeling the trunks of the trees—very carefully. Don does not remember it that way, but I like my story better than his. We did get seeds on the trip.

On another trip with Willard Yates we made a detailed search around Emison, where McCoy had collected his plant. We found nothing with pods as extreme as those on his specimen, but it was evident from this and other trips that there is tremendous variation in growth habit, spines, and pods of the honey locust. I have never tried to analyze the variation in any scientific fashion, and indeed it offers considerable difficulty. For one thing, there may be considerable variation of pods on a single tree as well as from tree to tree, and also since the pods are often borne high in the tree, it is difficult to get them unless one waits until they fall. However, from my observations I would say that the trees of the river bottoms in Indiana tend to be tall, moderately armed, and with rather straight and narrow pods, whereas the trees in upland sites are somewhat shorter, stoutly armed, and with broader, strongly twisted pods. The differences in the growth habit may depend to some extent on the habitat, but I think that the difference in pods is probably genetic.

It soon became apparent to me that there was no justification for accepting McCoy's new species, and later others, Donald Gordon and Duane Isley, the latter a specialist on legume systematics, came to the same conclusion. By now I had become most interested in locusts, so as time from my other studies permitted, I continued my investigation. It was more a hobby than anything else, and I have never published any of my observations. However, I feel that some of them are worth recording here, and perhaps they will inspire someone else to carry out a more detailed study.

Although Deam had recorded the hybrid *Gleditsia texana* for Indiana, I couldn't find it in any of the localities listed by him, but in 1959 on a trip with Rolla Tryon to Gibson County, we found a single specimen. The hybrid has since been found in Indiana by others, but I think that on the whole it is very rare, even allowing for the fact that staminate plants would likely be overlooked. One factor—but probably not the only one—that contributes to its rarity is that the two parent species usually flower at different times. For the last fifteen years I have kept dates on the flowering of some forty trees in my garden, and there has always been a two-week gap between the cessation of flowering of the honey locust and the opening of the flowers of the water locust. The tree that Rolla and I found was fruiting abundantly, and later tests showed that the seeds germinated.

Although *Gleditsia texana,* was originally thought to be a species, the evidence that it is a hybrid is very convincing. First, it is found only where the presumed parents occur together. Second, it is usually represented by a single tree instead of a population. Third, it is intermediate morphologically between the putative parents. Fourth, it does not breed true. Finally, artificial hybrids between the species agree in all particulars with *Gleditsia texana* in nature. One may ask how I could make artificial hybrids when my plants did not overlap in flowering. It so happens that after a long search I found one staminate honey locust on the outskirts of Bloomington that overlapped the flow-

ering of the water locust in the garden, and it furnished
the pollen for making the hybrids.

Since the hybrid is fertile and may overlap one or both
of the parent species in its flowering, one would expect
that backcrossing of the hybrids with the parent species
would occur. Thus, genes of the one species then might
pass into the other through backcrossing of the hybrids,
and in time one might expect an overlap in the flowering
of the parent species. However, this has not happened.
The reason for it is not known, but I am inclined to think it
the result of ecological factors. *Gleditsia aquatica* is adapted
to wet habitats and *G. triacanthos* to a variety of drier habitats where the hybrid and its derivatives would not be in
competition with the parent species, and they most likely
would be at a competitive disadvantage, for with their
mixture of genes they would not be as well adapted to
either a dry or wet habitat as are the parents.

On the other hand, it is difficult to deny that there
may have, at times, been some exchange of genes between
the two species. This could help explain the great variability in the honey locust, as well as the rare occurrence
of such plants as the one found by McCoy. The features
in which his plant differs from the honey locust—shorter,
straighter, dry seed pod—could all have been derived
from the water locust. The water locust has much less
variability than does the honey locust, perhaps in part
because it grows in a more nearly constant habitat, but
occasionally some trees are found with two-seeded pods
instead of the characteristic one. I find that when seeds
from the hybrid are grown in the garden, they often produce plants with two-seeded pods, probably as a result of
backcrossing to the water locust.

In 1960 and 1961 and again in 1972 I planted a large
number of locust trees, grown from seed from a variety
of sources, in a little-used part of our botany experimental
field at Indiana University. This garden has been a valuable source of information. Moreover, it has provided a
shady area at the field where one can rest after hoeing

Fruits of honey locust (*left*), water locust (*right*), and hybrid (*center*).

sunflowers. It has also become something of a bird sanctuary, and, although someone once told me that squirrels would not climb a honey locust because of the thorns, I have seen squirrels in the trees.

In the descriptions of *Gleditsia* in various floras I have found some disagreement about the nature of the sex of the trees.[2] Sometimes they are described as polygamous —with perfect and unisexual flowers on the same or on different individuals of the same species—or polygamo-dioecious—polygamous but chiefly having either staminate ("male") or pistillate ("female") flowers on a single individual. I have found that all the trees in the garden are strictly dioecious—having the staminate and pistillate

[2]Strictly speaking, as botanists know, the trees are sporophytes and hence do not have sex.

flowers on separate individuals. The pistillate flowers may have well-developed anthers, but I have never found them to produce pollen. Nor do the trees change sex, as has been suggested.

As already pointed out, one of the most conspicuous features of locusts is their thorns, and one might ask what is their function. The answer usually given for the presence of spines is that they are for protection, but protection from what? Perhaps a few hundred thousand years ago the spines protected the foliage from being eaten by some animal, but if so, that animal is no longer with us. Spineless forms are occasionally found in nature, and insofar as I can determine, they survive very well without spines. One spring I found that a number of my young honey locusts had been chewed off by some animal, and I think the culprit was a rabbit. Perhaps, I reasoned, the spines serve their function of protection when the plants are young. So I tried an experiment. I planted a dozen spineless and a dozen spiny plants at the field and tabulated the results the next spring. All of them were chewed off. In fact, the spines may actually serve the rabbits, for I have a little dog that delights in chasing rabbits and on rare occasions even catches one. Twice he has torn his ear on the spines, so badly that I had to take him to the vet to get the bleeding stopped.

Although I have never conducted any chemical tests on the two species, I think that I have good evidence that there are chemical differences between them. First of all, the seeds of the honey locust are nearly always attacked by bruchid beetles and those of water locust are not. In fact, in collecting last year's pods off the ground, or in the pods stored in my office for several months, I don't recall many pods of honey locust that did not have the neat little perforations from which the beetle escapes. I find that they also perforate the paper sacks in which pods are stored.

Although Sargent and others have stated that the honey locust is subject to few diseases, from 1972 to 1974 the honey locusts in much of Indiana suffered severe attacks from insects that left the leaves badly disfigured. The attacks were so bad that many of my trees failed to flower, and those pistillate ones that did flower, set very few fruits. Plants of the water locust, however, were not damaged.

A few years after the garden was started, it became apparent that the water locusts were doing as well as or better than the honey locusts. My garden was in an upland site where one might expect to find the latter occurring naturally, and hence I had expected them to outgrow the water locusts, which in nature are found only in or near water. So I became intrigued in trying to explain why the trees grow where they do naturally. I wondered if the taxonomic differences between the species might have any adaptive significance in this regard. The chief differences between the species, you may recall, are in the fruits and seeds. The fruits of the water locust are dry and one-seeded, and the seeds are rather flat and ovoid, whereas the pods of honey locust contain several seeds and a sweetish pulp, and the seeds are bean-shaped. The water locust produces many more pods than does the honey locust, but Greg Anderson estimated that the seed yield of each species is probably similar, since each pod of the latter contains many seeds.

Most of the fruits of the water locust end up beneath the tree, although some may be blown a short distance, perhaps at times up to two hundred feet. Donald Gordon has postulated that water is the chief means of dispersal, and I found that pods will float for up to two weeks in water trays in the greenhouse.

Sargent in 1890 stated that "the pods [of honey locust] contract in drying with a number of cork-screw twists and without this provision they would remain where they fall under the trees, but the pods thus twisted roll like wheels. Being very light they are blown for great distances over the frozen ground and especially over the snow. The obstacles they are obliged to overcome in their journeys probably help to break open the pods and liberate the seeds." I don't know if Sargent had actually observed this, and I thought it might be a flight of fancy, but Rolla Tryon has told me that he observed pods rolling over the frozen ground in Massachusetts. However, I hardly think that is the major means of dispersal. Water is probably also involved at times, for I have found that pods of this species will also float up to two weeks. Animals are probably the chief dispersal agents, and, as I have already mentioned, animals, particularly cattle, are known to eat the pods. Cattle, of course, could not have been responsible for dispersal until fairly recently, so perhaps deer were once the chief agents of dispersal. It has been suggested that passage of the seeds through the digestive tract of animals would enhance their germination. I thought of looking into this, but I could never bring myself to ask for a cow on a National Science Foundation grant application.[3]

The honey locust appears to have several rather effec-

---

[3] When I delivered a talk on honey locusts before the Botanical Society of America in 1981, John Beaman, then Program Director of Systematic Botany for the National Science Foundation, was in the audience. A few days after my talk I received the following letter from him:

In response to the request in the course of your eloquent presidential address to the Botanical Society of America, the Systematic Biology Program hereby authorizes transfer of funds in your grant DEB 80-22777 from the summer salary category to materials and supplies for the purchase of one Brahman bull. We trust that the product issuing from this purchase will provide the highest possible yield of processed honey locust seeds. If necessary, we shall also be pleased to authorize purchase, with any remaining summer salary funds, of honey pots for collection of said product.
Yours for a fertile investigation.

After a decent interval I replied as follows:

Please excuse my delay in replying to your letter of August 21, but I thought that I should wait until I had something to report. We now have our animal, a magnificent beast as you can see by the enclosed photograph. I want to thank you for the authorization that made this purchase possible.
There have been, however, a few problems. I find that "Sunflower" (as the students named him) can not subsist on a diet of honey locust pods alone. After he had consumed several of my perennial sunflowers and most of the vegetable gardens of two of the graduate students, I realized that we would have to provide

tive means of dispersal, which may in part account for the wide distribution of this species. Human beings obviously have enlarged the distribution in recent times. To return to water locust—could it be confined solely to wet areas because of its limited means of dispersal? It is hard to imagine that the seeds have not been carried to drier areas on many occasions.

Another possibility that deserved investigation was whether the water requirements for germination of the seeds of the two species differed. My plants in the garden had been started from seeds planted in the greenhouse, where they received water every day. Perhaps in nature the seeds of water locust require more water than those of the honey locust, which would be an explanation why the water locust was found only in wet places in nature.

I learned a lot about the seeds. I found that if nearly mature seeds were extracted from the pods, they would germinate immediately, but when seeds were collected from the pods found on the ground in the winter or the next spring and planted in the greenhouse, only very rarely would a single seed germinate even after a year or longer. This phenomenon of dormancy in seeds is well known in many plants. By filing a line through the seed coat until I could see the cotyledons, I found I could get almost 100 percent germination. This suggested that the

---

supplemental feed. I am surprised at how expensive it is. I am hoping that I can use some of the supplies allocation from my grant for this purpose. I realize, of course, that I shall have to get approval from the current Program Director before doing so.

In your letter you made no mention of how the animal should be disposed of at the conclusion of the experiment. Since the funds for the purchase came from my summer salary, I feel that I should be allowed to keep the animal. I find that I can get a meat locker locally. (I shall, of course, personally pay for the rental of the meat locker and the butchering costs.) I assume that I also need the approval of the current Program Director before dispatching "Sunflower."

The photograph that I sent him was taken in Ecuador. I should also mention that I never have held a research grant for the study of locusts. The grant referred to by John was for a study of loofah gourds.

seed coat was a barrier that did not allow water to enter the seed. Some people have treated the seeds with sulfuric acid for several hours and have found that the seeds then germinate rapidly. In nature, probably various agents — for example, freezing and thawing, or, as suggested earlier, passage through an animal's gut — may enhance germination.

Seeds of water locust gave very good germination up to five years, after which time germination percentages fell off rapidly. On the other hand, seeds of honey locust were found to give very good germination after ten years. Seeds older than that were not tested. This difference probably stems from the difference in the thickness of the seed coats of the two species, which I discovered when I was filing the seeds. Seed coats of the water locust averaged 0.87 mm in thickness, whereas those of the honey locust averaged 1.18 mm.

A number of experiments involving germination of the seeds under various moisture regimes was carried out, and no differences were found. In fact, I was surprised how little water was required for germination of the seeds of both species. I wondered if the young seedlings of the two species had different water requirements, so a number of experiments were attempted in the greenhouse, and again I could detect no difference between the species.

I also studied root growth by planting seeds in glasses. In soil with average moisture content, in twelve hours roots of the honey locust grew to lengths of more than 12 mm, whereas those of the water locust grew to only 10 mm, a difference that proved highly significant statistically. When seeds were planted in a saturated medium, the roots of the water locust outgrew those of the honey locust 4.5 mm to 3.3 mm in ten hours. At last I had found some difference, and as we shall see shortly, the difference in root growth might have some bearing on the success of the plants in nature. My experiments were hardly conclusive, and more experiments should be designed, particularly to see if they hold up in nature.

I also grew eight one-year-old seedlings of each species in pots in trays of water. After three years all the plants of the water locust were taller than those of the honey locust, and at the end of five years all the plants of the honey locust were dead, whereas those of the water locust were still thriving. These results were hardly a surprise and indicated why the honey locust would not be found growing in swamps, but it did not, of course, explain why the water locust had to grow in wet places.

Apparently there was only one explanation left to account for this—competition. This reason had occurred to me quite early, but for the sake of this story I have saved it until last. The ability, or lack of ability, of plants of a species to compete with those of other species is an important factor in controlling the distribution of plants. The bald cypress often does quite well when planted away from water, but in nature it is always found in or near water. It apparently cannot compete with other plants when it is not protected, or at least started, by human beings away from very wet habitats. My water locusts planted in the garden did not have to face competition with other plants, so perhaps competition was the answer.

My problem now was how to test this idea. I would have liked to have set up some study plots in areas where water locust grew naturally, but the nearest spot was over one hundred miles away, and I didn't see how I would arrange the frequent visits that would be required. The best I could come up with was to set up an experiment in our garden. In late 1976 I had an area plowed, about 46 by 48 yards, adjacent to my locust trees. Over the years I had observed pods of both honey locust and water locust in that area, so I thought there should be sufficient seed on or in the ground to provide a large number of plants of both species. The seedlings would be identified by species and marked with stakes so that their growth could be observed over a period of several years. The identification of seedlings of plants by species is not always easy, but the honey locust and water locust can readily be distinguished by their cotyledons (or "seed-leaves"), for these parallel the shape of the seeds—those of the water locust are ovate, whereas those of the honey locust are more lanceolate. The following spring 148 seedlings appeared, but unfortunately only three of them were of the honey locust; probably the reason that more did not appear was that my trees of honey locust had produced very few fruits the previous three years because of insect damage. A mole soon burrowed under one of these few seedlings, and it was lost. Seventy of the 148 plants of the water locust survived the first year.

Twenty-one plants of the water locust and the two honey locusts were still alive in the spring of 1982. I had more or less expected all of the water locusts to die very shortly, but some of them are presently over seven feet tall. The garden is now a mass of vegetation, and I have kept a list of all the species that have appeared, but I am going to mention only one of them. A large number of silver maples germinated in one corner of the plot, and these form a dense thicket, with some of the plants being over fifteen feet tall. Under their shade not a single locust has survived. I do not want to suggest that it is competition with silver maple in nature that limits the distribution of water locust, for many other species of plants may also be involved, but it may be of some significance that the distribution of silver maple largely overlaps that of water locust in nature except in Florida.

After I discovered that I had very poor germination of honey locust in the previous plot, I established another one the next year in which I scattered fifty seeds of each of the locusts over the plowed ground. Sixteen plants of the water locust and fifteen of the honey locust germinated the next spring. This plot is on slightly higher and drier ground than the other one, and the weed growth, particularly of woody plants, has not been as abundant as in the other plot. Ten of the plants of water locust and twelve of the honey locust were still alive in 1981, and there was a striking difference in the height of the two.

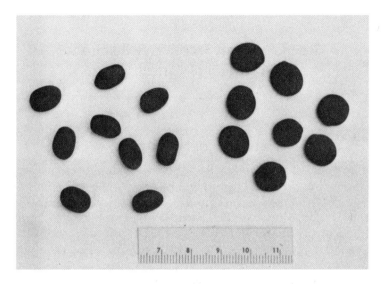

Seeds of honey locust (*left*) and water locust.

The water locusts averaged only slightly over one foot tall, whereas the honey locusts averaged three feet tall. These results were more or less what I expected, but exactly why the honey locust outgrows the other species is not known. However, the little experiment mentioned earlier with roots may be significant, and the more rapid root growth of the honey locust may explain why it grows better than does the water locust in the competition experiments and may well explain why the latter does not compete in dry sites in nature. If the hypothesis is correct that failure to compete is the reason water locust cannot grow away from water in nature, then the trees in my garden should never reach the fruiting stage; or, in the event any of them do (after all, their germination occurred on newly plowed ground), I would not expect them to reproduce themselves in the garden. I hope to continue my monitoring of them for some years to come.

CHAPTER 13

Seeds, Sex, and Sacrifice: Religion and the Origin of Agriculture

THE RAPE OF PERSEPHONE

The song of Demeter, whose beautiful daughter, Persephone, was violently seized by Hades with the collusion of her father, Zeus, is one of the most beautiful and profound of all Greek myths. Far away from her mother and her fruits of grain, Persephone was playing with the full-bosomed daughters of Oceanus, gathering flowers in a meadow. Persephone reached out her hand to pluck the lovely narcissus when the earth yawned wide, and from it leapt out Hades, who seized her and bore her away on his golden chariot. She screamed and called out for her father, but no one listened to her except tender-hearted Hecate. The peaks of the mountains and the depths of the sea rang with her voice, and her mother heard it. A sharp pain penetrated her heart, and she tore off her headdress and her cloak and sped like a bird over land and sea, searching for her daughter. No one, neither gods nor mortals, wanted to tell her the truth. For days Demeter wandered over the earth. Not once did she take ambrosia or nectar, for she was grieving. On the tenth day Hecate found her and told her that Persephone had been carried off, but she knew not by whom. Together they came to Helios, the watchman of gods and man, and the goddess asked him to tell her of her child. The son of Hyperion answered, "Queen Demeter, no one of the gods is guilty except Zeus, who gave her to his own brother, Hades, to be his blooming wife. Seizing her, Hades took her screaming down into the misty darkness."

Demeter was angry, and, staying far away from the gods, she disguised herself and went to the cities of men. She sat down at the Virgin's Well by the side of the road. The daughters of Celeus, son of Eleusis, saw her when they came for water. Not recognizing her, they asked who she was and where she came from. She told them that her name was Doso, that pirates had carried her away from Crete, and that she had fled from them when they reached Attica. She then asked to whose house she might go to find work. Returning home, the daughters told their mother, who had a newborn son, and she sent them back to bring Demeter to raise her child. So Demeter came to them, but still grieving over her lost daughter, she did not partake of food or wine, only barley water. Demeter took the child onto her lap and promised to nurse him. The son grew up like a god under Demeter's care, and at night she hid him from his parents in the heart of the fire. Demeter would have made him immortal, but one night the mother spied her and cried out. Demeter heard her and cast the infant to the ground. "Foolish ones," she said, "I would have made the child immortal, but now there is no way in which he can escape death, for I am Demeter. Let the people build me a great temple." As she spoke, the old age left her and she was once again beautiful.

When the morning came, the women told Celeus what had happened, and he ordered a great temple built to Demeter. When the temple was finished, according to the will of the goddess, the laborers went home, but Demeter remained there. Wasting away and yearning for her daughter, she fashioned a terrible year for mankind. The ground would not let the seed sprout, for Demeter kept it hidden, and much barley fell on the ground in vain. She would have destroyed the whole of humankind with famine and deprived the gods of their glorious gifts and honors had not Zeus seen it. So first he sent Iris to her. She came to Eleusis and told Demeter, "Father Zeus calls you to the families of the gods." She begged her to come, but Demeter was not persuaded. So then Zeus sent all of the gods, who pleaded with her, but still she rejected them, saying that she would never set foot on Olympus or cause fruit to come from the earth until she saw her fair daughter. Hearing this, Zeus sent Hermes to persuade Hades to allow Persephone to come from the dark to the light so that her mother, on seeing her, would put aside her anger. When he reached the underworld, Hermes relayed Zeus's message, so Hades called his unwilling wife, Persephone, and told her to return to her mother. At this, Persephone was happy, but Hades gave her a pomegranate seed to eat in secret, lest she remain at the side of Demeter for all time.

When Hermes reached the temple with Persephone, she raced to meet Demeter, and they embraced. While holding her child, Demeter sensed a trick and asked her if she had taken any food while she was down below. If she had, Demeter told her, she must return to the secret part of the world for a third part of the year, but then she would return to the earth each year when it bloomed with flowers. Persephone then told her mother that Hades had forced her to eat a seed of pomegranate against her will.

So it was to be, Zeus agreed, and Demeter once again made the land fertile with grain. And going to the kings and leaders of man, she taught them the sacred rites.

This condensed account of Demeter and Persephone is only superficially the story of a mother losing her daughter; in reality it is a story of the corn that "dies" each year and is born again the next.[1] Persephone, or Core in some accounts, is the grain that goes underground and must always return there for a portion of the year. Demeter, who was to become Ceres in Rome, was a goddess of vegetation. It was she who gave corn to humankind and taught Triptolemus the art of growing it. It may be that Persephone and Demeter are one and the same—Persephone representing the young grain and Demeter the mature crop. Demeter's role as a fertility goddess is seen in another account in which she lay with Iasion in a "thriceplowed" field, for sexual intercourse of the gods was once thought to be essential for the fertility of the fields.

---

[1]Corn is a common generic term for grain or cereal. In the Near East it is barley or wheat, whereas in North America it is maize.

The date at which the hymn to Demeter was written is unknown, but it is thought to have come from the first millennium before Christ, which, of course, is several thousands of years after agriculture was known in Greece. Possibly the myth came to Greece along with agriculture, at which time it was probably quite different from, and much simpler than, the version that has come down to us. The myth itself tells us that Demeter came from Crete, and there is also some external evidence suggesting this origin. Beyond that we can only guess. It could have come to Crete, if that indeed is its source, from the Near East, with Demeter being another name for one of the early vegetation goddesses of that region. Although Frazer argues against Demeter being an earth goddess, Paul Friedrich recently pointed out that one of the more plausible etymologies of her name is "earth mother." Although I have no intention of going into a detailed discussion of the myth, which has already received so much attention from others, two points should be made. Demeter does not give grain to humans in this account; rather, she withholds it. An "original" myth would have her giving corn to man. Also we see that Persephone has to go underground for a third of each year. At the time of the writing of the myth, Greek farmers were planting seed in the spring so that the grain would be underground for only a few days before it germinated. Thus one might argue that the myth goes back to a time when planting immediately followed the harvest.

The cult of Demeter was celebrated at Eleusis for hundreds of years, traces still remaining in the beginning of the nineteenth century. It probably became more than an agricultural ritual, perhaps encompassing the death and rebirth of humans as well. What transpired at the festival and the identity of the drink *kykeon,* the barley water of the hymn, still intrigue our imagination, and both have been the subjects of books in recent years.

Although I saw no clues in the story of Demeter, as I read it I wondered if earlier myths or religious practices might reveal anything of the origin of planting or agriculture. Certainly primitive religion was closely associated with agriculture; we can look to agriculture as a prominent influence on the development of religion, and possibly there was a reciprocal relationship. Before examining the subject further, I should like to say something about one of my chief sources of information, Sir James G. Frazer's *The Golden Bough.*

FRAZER

James George Frazer (born 1854, Glasgow; died 1941, Cambridge), classical scholar, folklorist, and anthropologist, brought out *The Golden Bough: A Study in Magic and Religion* in two volumes in 1890. It was expanded to twelve volumes in 1915. Although *The Golden Bough* has become a classic, it has been the subject of considerable criticism, some of it rather vicious, in recent years. Many, if not most, of Frazer's theories are not accepted today as a result of modern research. Some of his "facts" have been questioned, particularly those that came from uncritical observations of travelers and missionaries. It has even been said that to achieve a well-turned phrase he would sometimes disregard the facts (he did write extremely well). In his supplement to *The Golden Bough, Aftermath,* published in 1937, he wrote, "My writing should not survive for its theories, but for the facts," and this is largely true. The focus and methods of anthropological research have changed over the years. Today most anthropologists regard fieldwork as a necessity, and Frazer was an "armchair" anthropologist whose research was carried out in the library.

The Golden Bough is still in print, attesting to its importance, as is an abridged edition by Frazer himself, but probably few people read these today. In fact, an anthropologist with whom I discussed the subject asked, "Has anyone read all of Frazer?" I certainly haven't, although I have consulted the unabridged edition, but the answer is "yes." Theodor H. Gaster has, and produced an abridge-

ment that still enjoys considerable popularity. His *New
Golden Bough* has been my chief source concerning myths
and customs connected with ancient and primitive agriculture. Gaster's work preserves the spirit of the original
while eliminating unreliable information and adding new
interpretations. In his editor's foreword he gives an interesting and fair evaluation of Frazer's work. Unfortunately,
I can't say the same of Downie's biography of Frazer, the
only one yet to appear.

Some thirty years ago when one of my colleagues suggested we order *The Golden Bough* for our biology library,
I was skeptical. Although I have used it extensively since
that time, I am still not sure that it belongs in the biology
library, and I am fairly certain that if it were there it
would receive little or no attention from the modern bi-
ologist (a pity, for many of them need broadening). Frazer's
work has inspired not only my efforts but those of others
as well—from Grant Allen, who wrote on the origin of
agriculture in 1894, to Thomas Tryon, who wrote *Harvest
Home*, a recent fictional account of a modern village in
the eastern United States in which many of the ancient
agricultural customs are preserved, including the killing
of the "Harvest Lord." Although the author does not ac-
knowledge Frazer, surely he was influenced by his works.

In reading Frazer's accounts of the rituals associated
with early agriculture, I wondered if he had ever come to
any conclusions regarding the origins of planting and agriculture. The subject does not seem to have been of particular interest to him, but he does make a few comments
from time to time. For example, in *The Golden Bough* he
states that in some of the customs of savages totally ignorant of agriculture perhaps can be detected some of
the steps by which people advanced to the cultivation of
plants. Digging the earth for roots may have served at
times to enrich the soil, thus attracting a larger number
of people and allowing them to subsist for a longer time
at the spot. Moreover, the winnowing of the seeds on the
ground that had been turned up with a digging stick would

foster the same result, for some of the seeds would have escaped and fallen on the soil and borne fruit. Although the aim of these people was nothing more than satisfying their immediate hunger, "the savage man or rather the savage woman was unconsciously preparing the people for a more abundant source of food which would allow them to increase and abandon their migratory ways. So curiously sometimes does man, aiming his shafts at a near but petty mark, hit a greater and more distant target." He goes on to point out that women more than men have contributed to the greatest advance in economic history—a settled life and an artificial subsistence, that is, agriculture.[2]

It is in his *Totemism and Exogamy* that Frazer makes his most interesting comments. In volume one, in his description of the elaborate ceremonies of the Kaitish of Australia to promote the growth of their food grass, we learn that the headman of the grass-seed totem takes some of the grass seed to the men's private camp and grinds it. One of the men there puts some of the seed in the headman's mouth, and "he blows it away in every direction, which is supposed to make the grass seed grow plentifully everywhere." At the conclusion of his discussion of the ceremony, Frazer points out that the blowing of the grass seed is a particularly interesting feature, for such a procedure might really have the intended effect of propagating the seed, and if the natives observed, as they might well have, the success of the ceremony, they might in time come to sow the seed without the accompaniment of those chants or spells to which at first they ascribed a great part of the efficacy of the rite.[3] Thus a rational agriculture might

---

[2] Frazer, however, was not the first to credit the origin of agriculture to women. Bachofen in 1861 stated that the observations of people living today have revealed that the impetus to agriculture is derived from women, whereas men tend to resist the change. Since his time many other writers have given the chief credit to women, and Erich Neuman tells us, in *The Great Mother: An Analysis of the Archetype* (2d ed., Princeton, 1973), that women domesticated not only plants but beasts and man as well. Some of those in the modern feminist movement might object

spring directly from what was originally a magical ceremony. "May not this, or something like it, have been in
more advanced communities the real origin of agriculture?" Frazer asks. In volume four of the same work he
again comments and points out that it might be argued
that totemism not only led to agriculture but to the domestication of animals as well; however, the Kaitish are
ignorant of the fact that a seed when planted in the ground
will grow. Agriculture, as he tells us in other places, of
course, never came to Australia until it was introduced
by Europeans.

THE GREAT MOTHER

Female figurines or "Venuses," made of stone, bone, or
clay, have been found in Europe and Asia in preagricultural sites, some of them going back to twenty thousand
years ago or earlier. In some of the figurines the female
organs are prominent while the facial features are scarcely
represented, and some appear to be in an advanced state
of pregnancy. Thus it has been thought that these idols
are somehow associated with fertility—not only of humans
but perhaps of other animals and the vegetation as well,
for the greatest concerns and mysteries of the people were
life, death, and food. From these have also come the idea
that humans' early religion involved an Earth Mother or
Earth Goddess or Mother Goddess, who brings forth all
life.

It must admitted, however, that the actual significance
of these figurines is unknown, and whatever explanation

---

to part of his statement, for they do not feel that she has, as yet, completely domesticated man.

[3] It is not entirely clear from the passage cited whether the headman
put ground or whole seeds in his mouth. If the former, of course, there
would be no possibility of the seeds germinating. However, we might
suppose at some time and place unbroken seeds may have been so
employed.

we may come up with may be quite different from the
actual one. It could also be that there is no single reason
for them. Perhaps they represent nothing more than man's
(assuming they were made by men and not women) pre-
occupation with sex, an early example of the *Playboy* men-
tality. In no way can it now be proved, however, that they
were not of a religious nature and somehow connected to
fertility.

It is probably of some significance that representations
of the phallus do not occur until much later in the archaeological record. Men apparently had finally discovered their role in reproduction, and as they did so, the
Great Mother Goddess eventually gave way to male gods
in early religions among agricultural societies. A goddess,
however, was still usually assigned to vegetation and fertility of the fields.[4] A figurine of a woman has been found
in a grain bin at Catal Hüjük, an early agricultural site
in Anatolia, and perhaps in her we can see the direct
descendant of the fertility Venuses of the preagricultural
era.

Today "mother earth" is largely a figure of speech, but
mother earth was at one time sacred, not only in Europe
and the Near East but in other parts of the world as well
—and still is in some places. A well-defined mother goddess cult, in which the people still made female idols with
the female parts of the anatomy emphasized, was known in
Nigeria in this century.

Who is to say whether the mother goddess cult developed in one place and then diffused around the world in
preagricultural times or whether it involved independently in many different places and at different times?
It seems most likely, however, that the early fertility goddesses, such as Astarte, Inanna, and Isis as well as Demeter,
of the agricultural people in the Near East and adjacent

[4]Not surprisingly, some of the recent feminist writers have developed
the thesis that the sexes were more nearly equal when female gods were
present and that the subjugation of women began when monotheism
replaced polytheism.

territories can be traced to a single earth goddess of pre-agricultural people.

MYTHS

Myths accounting for the origins of food plants are widespread among archaic people just as are myths to account for the origins of the people themselves. In fact, many myths of the origins of humankind include the origins of their foods as well, and in many such myths the people themselves come from plants. One must assume that it was extremely important to the people to explain their origins, and to them their myths offered satisfaction. Yet for us to interpret these myths is no simple matter. To define, explain, and understand species is far easier than to do the same for myths. According to one school, myths are like dreams, dreams of a society instead of an individual, and according to Freud, myths and dreams work in the same way. They come from the imagination, not reality, but at the same time they are very real to the people concerned. Myths have been said to be a practical response to the unknown and an attempt to control nature. Some have thought that fertility myths and their accompanying ritual were magical in nature, intended to provide for continuity and to ensure good harvests in the future. We may find myths full of contradictions and absurdities and meaningless. Yet in the hands of an expert their analysis may reveal "fundamental truths," one "expert" tells us, while another one tells us that any explanation today will falsify such a myth by giving it a logic it never had.

The song of Demeter is not typical of myths explaining the origin of food plants. In the Near East, for example, it is usually a vegetation god, Dumuzi or Baal, who must go underground. Most myths of food origins from other parts of the world are much simpler than Persephone's story and often very crude. For example, a myth of the Caingangs of Brazil tells us that when their ancestors suffered from a lack of food, their chief told them to till a piece of land by fastening a vine around his neck and dragging him over the ground, then a few months later they found that his penis had produced maize; his testicles, beans; and his head, gourds. Having gourds come from the head shows a certain logic, and in myths from southeast Asia, coconuts spring from the heads of people. In some myths it is human teeth that give rise to maize. Some of the agricultural myths may seem like fairy tales, but myths may be distinguished from other forms of folklore by their sacred nature.

Myths dealing with the origin of food plants take many forms. One of the simplest is that the plants originate from a sexual union of the female earth with the male heaven, the rain being the fertilizing agent. The plants may be given to the people by the gods, or they may be stolen from the gods, often by being concealed in the vagina or the penis. If not acquired from the gods, they may come from a culture hero or from some other human in some extraordinary way. Sometimes it is an animal or tree that is responsible.[5] A very frequent theme is that the plants originated from the corpse of a god, an ancestor (usually female), or a culture hero. With such great diversity, it is difficult to see a single origin for all agricultural myths. On the other hand, some of the myths from the Americas and southeastern Asia are so similar that it is difficult not to ascribe to them a single origin.

Most of the myths telling of the origin of plants deal with domesticated plants. This is perhaps not surprising. There were few hunters and gatherers left when myths began to be collected. So it is difficult to say how widespread myths were that dealt with the origin of the major

[5]There is a huge body of myth and ritual dealing with trees, and tree worship at one time was widespread. The Tree of Life or the Tree of Knowledge is common in many cultures. Trees or groves of trees were regarded as sacred in Europe until fairly recently. Trees, such as the coconut, it should be noted, may have been the chief source of food of some people, but not all tree worship can be explained on the basis of food alone.

wild food sources, though we know they existed. For example, we find that nonagricultural Indians of California, who depended on the acorn for food, had a culture hero from whose ashes the oak appeared.

Did myths at times impede the development of agriculture? The Menominees of the Great Lakes region did not sow wild rice because their myth told them they would always have it, and some western North American Indians were forbidden to plant. In reply to the question why he did not do so, one Indian replied, "You ask me to plow the ground. Shall I take a knife and tear my mother's bosom?" Mother earth was indeed sacred. One cannot help but wonder, however, if myths prohibiting planting arose only after some people learned that their neighbors practiced agriculture and felt that it was necessary to explain why they themselves did not. Surely, if their livelihood had demanded it, the people would have changed their myth to accommodate the change.

Myths, we know, may change with time for various reasons, which helps to explain some of the absurdities. For example, parts of myths of conquerors or the conquered may be incorporated in another people's myths. As new plants were acquired, they were added to the older myths. For example, the watermelon is mentioned in the creation myths of some southwestern North American Indians, and we know that the watermelon was not present in the Americas until after the arrival of the Spanish. Maize is frequently found in the myths of southeastern Asia. We know that it also must be a late addition, for maize did not reach Asia before the sixteenth century. Although some hold that it reached Asia in pre-Columbian times, there is no good evidence to support that contention. The myths themselves can hardly be used to attempt to prove otherwise.

Myths explaining the origins of food plants were probably universal among early agriculturists and, of course, persist in many cultures today. Can they tell us anything about the origin of agriculture that would be scientifically

acceptable? One of the best such attempts is that of Adolf
E. Jensen, who made an analysis of many such myths.
He finds that in general a distinction can be made between the myths of seed cultivators, who stole the cereals
from the gods, according to his Prometheus mythologem;
and the myths of vegetative cultivators, whose tuber crops
came from the body of a murdered deity, according to his
Hainwele mythologem. His analysis suggests to him that
the latter theme is older, and this in turn suggests a tropical hearth for agriculture in southeastern Asia. The archaeological record provides no support for his hypothesis, however, as will be shown below.

SEX AND SOWING

The principle of agriculture is the principle of ordered sexual
union. . . . It is not the earth that imitates women, but women
who imitate the earth. [Bachofen, 1861]

If we accept that early peoples believed that the marriage
of the gods was responsible for all creation and that the
fertility of the earth and of human beings were related,
it should come as no surprise to learn that the human sex
act came to play a role in the fertility of the fields. In
the Near East and adjacent areas the king or queen, as the
representative of the gods, was the main participant in the
fertility ritual, and in many parts of the world the people
engaged in a ritual mass mating in imitation of the gods to
assure the fertility of the crops. It is reported that among
the Pipiles of Central America certain persons were ap-
pointed to carry out intercourse at the very moment the
seeds were planted.

While perhaps we should allow for some exaggeration
in some of the reports, others leave a great deal to our
imagination. Particularly from some of the Victorian writers in the last century we learn only that festivals of "sexual
looseness" of a "licentious character," "disgusting rites,"
and "sexual orgies" took place, and we are spared the

details. In most societies such practices took place only at planting, but in others they are also known from harvest times, and in a few apparently every festival was so celebrated. Some of the people who carried out such rites at the time of planting were straitlaced about sex at all other times. In other societies the sex act was forbidden at the time of planting, and sexual offenses were thought to bring infertility to the fields. Thus, in one way or another, we see that human sex was thought to be closely associated with the fertility of the crops. In fact, food and sex are often thought to be closely related; for example, among the Kogi of Colombia the act of eating symbolizes the sex act.

Some of the rites associated with planting or harvest were carried to such extremes that they came to be suppressed by the officials in both Greece and Rome. The temple prostitutes of the Near East developed from the older sacred marriages, and they were condemned by the Israelites, who were descendants of pastoral, not agricultural, people and hence did not understand the association of the prostitutes with fertility. Thus the sacred prostitution of that region became fornication and adultery to followers of Yahweh.

Although it was usually the generative powers of women that were thought to be of the greatest importance to fertility, the men's role also became appreciated. Phallic worship developed in some cultures and persisted in Africa until very recently, and it may still be practiced in some places. In ancient Egypt women carried phallic images operated by strings in the cult of Osiris, who was a fertility god. Frazer tells us that "in the Chambers dedicated to [Osiris] at Philae the dead god is portrayed lying on his bier in an attitude which indicated in the plainest way that even in death his generative virtue was not extinct but only suspended, ready to prove a source of life and fertility to the world when the opportunity should offer." How true, for after his death Osiris in union with Isis conceived Horus. In India until this century the Maharaja of Patiala had to appear naked once a year before his people, "his organ in full and glorious erection."

In all probability, phallic worship developed only after man's contribution to paternity was appreciated. It has been suggested that this came with the development of agriculture, man's role being understood only after domesticated animals had been acquired. The replacement of female gods by male gods can also be seen as developing with recognition of man's significance in reproduction, as was mentioned previously.

The plow itself became a phallic symbol, for with it man penetrated the mother earth and opened the furrow. A Sumerian poet has the goddess Inanna comparing her vulva to the field and asking who will plow her vulva for her. A ritual plowing was to develop not only in Greece but also in India and China, probably having some significance for fertility. On the other hand, in *The Second Sex,* Simone de Beauvoire sees in the equation of the plow and the phallus a symbol of male authority and the subjugation of women.

Did reenactment of the sacred marriage take place in preagricultural times? I have found no such references among seed collectors. If man's role in paternity was not recognized until after the beginning of agriculture, it is possible that human copulation to promote fertility arose at the same time. Thus it is difficult to visualize human sex as having anything to do with the origin of planting unless one supposes that people somehow learned to plant seeds of plants in mother earth just as man planted his seed in woman, which would indicate a fairly sophisticated knowledge of sex in preagricultural times. Certainly the similarity of the two kinds of seeds was early recognized, for we find the same word used for both in ancient languages just as today we sometimes speak of man's seeds. One of the Mochica gods of Peru is depicted on pottery as casting his seeds on the ground, and a myth from New Guinea has the culture hero producing food plants by sowing his own semen. It seems a little far-fetched, however, to see people sowing seeds of plants as a result of their own sex act.

As with many other aspects of primitive agriculture, sex rites to promote fertility are known from both the Old World and the New, and we may inquire if they had but a single origin or arose independently. If such rites existed in the Old World long before agriculture began, they may have come to the Americas very early with people who did not yet have agriculture; they could have come to the Americas along with people who introduced agriculture (if such occurred); or they could have come to the Americas after agriculture was already known but long before the arrival of the Spanish. If one prefers to see the sex rites arising independently, one can assume that the idea that there was a relationship between human sex and fertility of the fields would have been fairly obvious to the primitive mind.

SACRIFICE

The sacrifice of human beings and animals seems to have been widely associated with early agriculture. Animal sacrifice is still practiced in some places, but the practice of human sacrifice seems unlikely today, although as recently as 1959 there was a report from Africa of a person being killed and the parts of his body being scattered on the field as was done in some of the old fertility rites.

Just as with myths, there are many questions concerning the meaning of sacrifices, and many explanations have been offered. One of the simplest is that the sacrifice is an offering to a deity. It was food for the gods, who, like mortals, needed food as well as sex. It was necessary to renew the energies of the gods, which had been exhausted in creating life and increasing fertility. It was an act to bring people in union with the gods. A human being was put to death as a representative of the life or spirit of the food plants; thus, life was offered to preserve life. Or it was a rite of renewal as a repetition of the act of creation. It is not known which one or ones of these reasons is or are correct, and they are not all mutually exclusive. There does not necessarily have to be a single reason.

The view that Jensen has elaborated—that it was a rite of renewal—has many appealing features. He sees the origin of death and the origin of food plants as being inseparable, as the original death of a god, from whose body the plants came, had given the people not only food but mortality as well. Therefore the ritual killing of a human being had to be repeated annually to maintain the fertility of the plants and also of the people themselves. Thus the killing was not to produce a gift for the gods; it was "a religiously founded ethical action" and a "festive reformulation of a primeval event." It was originally a ritual killing, not a sacrifice, and was only later to become a sacrifice.

Neuman saw blood as the essential element in sacrifice to the mother goddess, for a sacrifice of blood was closely bound to the decisive moments in the life of a woman— "menstruation, deflowering, conception and child bearing." The importance of blood in fertility rites continues to the present, for among the Ashantis of West Africa blood of a fowl is allowed to drip on the earth before planting, and the Aymaras of Bolivia sprinkle blood of a llama upon the seed potatoes before planting them.

In one of the hypotheses of the origin of agriculture put forth in the last century, Grant Allen called upon sacrifice to play the fundamental role. He believed that people observed that plants would grow exceptionally well on a newly made grave. From this, he concluded, the people would reason that the human burial was responsible, and this would have led to an annual sacrifice and burial in order to achieve a good harvest.

Recently another hypothesis has been proposed by Eric Isaac, who sees a religious motivation for the origin of planting. He postulates a ritual enactment of the primeval event, the killing of the god, leading to a ritual "killing" of plants by cutting them up and burying them. Such an act, of course, would lead to the vegetative propagation of root and tuber crops. His hypothesis would suppose that the sacrifice of the plant victim would have preceded the innovation of planting. Although nearly all of the examples of human sacrifice are found among people who had agriculture, we might suppose that some gathering people practiced the ritual killing before the advent of planting and agriculture.

While I feel that Isaac's hypothesis deserves serious consideration as an explanation of the origin of vegetative cultivation, it does not explain seed planting, and for its origin we must turn to a different kind of sacrifice.

FIRST FRUITS AND SACRED GARDENS

The offerings of the first fruits of the harvest to the gods may in a broad sense be considered a sacrifice, but these offerings stand apart from other sacrifices and are perhaps better termed a propitiation. The first fruits were offered to Demeter in Greece. Similar offerings were made in many places in both the Old and New World, and ceremonies of first fruits are still carried out in parts of Africa. These ceremonies or festivals usually involved the most basic of the foods, grains among seed farmers and the principal root or tuber crop among vegetative propagators. The offerings were made in various ways—often burnt or cooked before they were returned to the gods, or eaten by the people themselves in a special feast—all of which I am inclined to consider as late developments. Probably more ancient were rites of scattering the freshly harvested seeds over the ground or actually burying them.

The people who practiced rites of first fruits regarded the plants as alive and pervaded by spirits, hence sacred. Lamentations of the reapers accompanied the harvesting of the grains in ancient Egypt as their sickles killed the corn god. The spirits or gods in the plant apparently had both a sinister aspect related to famine and death and a

life-giving and fertilizing aspect. Thus the return of the
first fruits would be a desacralization to allow the people
to eat the rest of the harvest safely and at the same time
to ensure that there would be bountiful harvests in future
years. There seems to be some general agreement on this
interpretation, but it has also been suggested that eating
of the first, or sacred, fruits would make the people godlike
themselves.

At times the last sheaf in the field may have functioned
in the same manner as the first fruits. Frazer attached
considerable significance to the last sheaf in European
cultures, basing his conclusions largely on the work of
Mannhardt. Mannhardt's study, however, was later se-
verely criticized by von Syndow, who concluded that the
ceremonies connected with the last sheaf were simply festive occasions, often in the nature of pranks, and not
survivals of older rites. In spite of his criticisms, however, I find it difficult not to see in some of these observances remnants of what may at one time have been a more
sacred ceremony, much in the nature of the rites of the
first fruits.

From Frazer we learn that some peoples who did not
have agriculture also regarded the first fruits of the season
as sacred. I am unable to add to what he tells us except
that the Menominees of the Great Lakes region who depended upon wild rice (*Zizania aquatica*) as their principal
food had a first-fruits celebration. We also know, of course,
that certain hunters regularly made an offering of parts
of their kills to the spirits.

Special plantings of seeds are known from early agricultural people in the Near East, Europe, Egypt, and
India to produce what Frazer calls "Gardens of Adonis"
after the fertility god who was dead and buried and then
returned. Various seeds were sown, often in pots and
tended by women, and Frazer thinks the planting was carried out as a magic rite or charm to promote the growth
of vegetation. In Egypt effigies of Osiris made of earth
and grain were buried, and when later dug up the grain

had sprouted. The Tartars of Persia planted seeds in a jar, saying that they did so in memory of the creation. Still today in places in southeastern Asia special sacred plantings of rice are made. In Sarawak some families have a sacred rice that differs from the common variety and is planted separately. The emperor of Japan annually makes a ritual planting of rice on the imperial grounds. In America somewhat similar rites or customs were known, and the kernels of the sacred corn often were the first planted the following year.

Whether such sacred plants or sacred gardens have anything to do with rites of first fruits is purely conjectural, but I think we may perhaps see some connection. Of all the religious activity that accompanied early agriculture, it is in the rites of first fruits and sacred gardens that I can see an origin of seed planting and subsequently agriculture. I shall develop this thesis at greater length, but here I should call attention to their possible significance with respect to changes in the plants themselves. Hatt cites Wirz in 1927 for pointing out that the choosing of special plants for the rice mother may have had a practical significance, and I can see that the rites of first seeds might have functioned as a selective agent in the development of domesticated plants if the seeds served for planting in the following year.

Domesticated plants usually differ from their wild relatives in a number of traits, and I had long puzzled over how people in early times were involved in these changes. Certain of them may well have occurred by unconscious selection. Thus we might explain the fact that many, if not most, domesticated plants have lost their natural ability for seed dispersal—for example, the cereals have a nonbrittle rachis so that the grains stay on the plant, in contrast to the brittle rachis of the wild plants that allows the grains to scatter at maturity. When human beings started cultivating plants, they would likely harvest all of the grains from plants with the nonbrittle rachis, whereas some grains from plants with the brittle rachis would escape them. Thus in time more and more of the seeds for planting would have come from the plants with the non-brittle rachis. It is difficult, however, not to see conscious selection being involved in some of the changes in the plants. If we suppose that sometimes plants with special characteristics were chosen for the harvest of the first fruits because they were considered specially divine — perhaps any plant that somehow stood out was considered sacred — then we can account for the conscious selection of certain seeds for planting. For this reason unusually large fruits or seeds might be saved for planting, not because the people thought they in turn would produce larger fruits or seeds but because they were more sacred than the others. If indeed the first fruits to ripen each year were selected for the rites and then used for planting, we might eventually have an earlier-maturing strain coming into existence. Probably special attention was paid to seeds of unusual colors, and we find quite often that the seed color of the domesticated plant is different from that of the wild type; light-colored seeds are found in some domesticates, whereas the wild type is frequently dark. This is all very speculative, of course, but we have the example of rice previously mentioned, and it is known, of course, that some American Indians saved special ears for the seed corn and that the seeds of the sacred corn were planted the next year. The conscious saving of some seed for planting because it was sacred, along with unconscious selection, could explain the change from the wild to the domesticated type. Not all of the changes, of course, would be of economic benefit to the cultivators, but some of them likely would have.

THE ORIGIN OF AGRICULTURE

To my mind, it is extraordinary that anyone could have thought it worthwhile to speculate about what might have been the origin of some custom or belief, when there is absolutely no means of discovering, in the absence of historical evidence, what was its origin. . . . I hold that it is not sound scientific method to seek for origins, especially when they cannot be found. [E. E. Evans-Pritchard, 1965]

Although Evans-Pritchard was writing about primitive religion, his statement may well apply to the origin of agriculture, but if so, it certainly has not had much influence in recent years. Up until near the middle of this century, scientists for the most part neglected the subject, probably largely because of a lack of interest, perhaps stemming from a belief that the human race naturally tried to improve itself, and agriculture was a logical and natural outgrowth of such an attempt. Today the situation is changed, and we have many hypotheses, but none that has gained general acceptance. Why should we seek the origin of agriculture? Perhaps the answer is that people are interested in origins, and after all, speculation involves little cost or effort and is a rather diverting pastime.

I do not intend to go into all of the hypotheses concerning the origin of agriculture (references will be given to many of them for the interested reader), but some of them require comment. I would be remiss if I didn't discuss the ideas of Carl O. Sauer, who was a geographer at the University of California, and Edgar Anderson, who was associated with the Missouri Botanical Garden, for both of them were professors of mine and undoubtedly turned my thoughts in this direction. Moreover, both of their hypotheses are still frequently mentioned today. Some years ago Professor Sauer concluded that agriculture most likely first developed among fishermen in southeastern Asia. He saw the development of agriculture as requiring considerable experimentation, and such people would have had a dependable source of food so that the time would have been available for experimentation. Moreover, in contrast to hunters and gatherers, they would likely have been sedentary most of the year, which would also have been important to people trying to grow plants. He believed that the earliest agriculture was with root and tuber crops, reasoning that the knowledge of their reproduction was easier to acquire than that of seed plants.

In 1952, Professor Anderson developed his "dump-heap" hypothesis, which did not necessarily contradict anything in Sauer's arguments. He saw refuse sites around the early settlements of humans serving as a breeding ground for weeds and cultivated plants, often the same species. Such places, being open-disturbed habitats and also probably rich in nitrogen, would have promoted the vigorous growth of seeds or other plant parts accidentally or intentionally thrown on them. In 1916, apparently unknown to Anderson, a hypothesis very similar to his was put forth by Th. H. Engelbrecht.

Neither Sauer's nor Anderson's hypothesis called upon demographic factors as having any role in the origin of agriculture. In fact, Sauer's was quite to the contrary, because he did not see necessity as the mother of invention as far as agriculture was concerned. In more recent years, however, population pressure has been seen by several scientists as the most likely explanation for the origin of agriculture. In a whole book devoted to the subject, the archaeologist Mark Cohen has developed the thesis that the only factor that can explain what he sees as the irreversible and nearly uniform emergence of agriculture throughout the world is the increase of the population beyond the size that hunting and gathering would support. The events leading to the development of agriculture in various parts of the world, Cohen maintains, show remarkable parallels. Over eleven thousand years ago hunters and gatherers had occupied all the lands that would support their life-styles, and they were forced to turn more and more to unpalatable foods. The people who started agriculture were not verging upon starvation; the population pressure was "nothing more than an imbalance between a population, its choices of food, and its work standards, which forced the population to change its eating habits or to work harder." Although agriculture did not provide a better diet, greater dietary reliability, nor greater ease in the quest for food, it did provide more calories per unit of time and per unit of space than could hunting and gathering.

Now for the facts, and the facts come from the archaeological record. The finding of plant remains that can be referred to domesticated plants—for example, seeds or fruits that are larger than the ones found in wild plants or ones that have lost their natural means of dispersal—is usually thought to indicate domesticated plants and hence agriculture or, at least, incipient agriculture. Recent archaeological evidence, much of it since the time of Sauer's work, has considerably changed some of our earlier ideas concerning the times and places of the beginnings of agriculture. The earliest evidence comes from the Near East, where we now have evidence for the cultivation of both wheat and barley at about 7000 B.C.; they were shortly followed by lentils and peas.

Although some people have recently maintained that agriculture is as early in Southeast Asia as it is in the Near East, I do not see the support from the archaeological record. Plant remains from Thailand dated at about 7000 B.C. have been found, but the plants are not well preserved, the identifications are not certain, and it is far from clear that they are domesticated plants. Rice, which was to become one of the staples of that area, is not definitely known until around 5000 B.C. However, it seems likely that there was a very early center of domestication in Southeast Asia, even though it is probably somewhat later than that in the Near East.

In the Americas the first evidences of domesticated plants come from Peru and Mexico. A fairly recent discovery of two species of beans and a *Capsicum* pepper in highland Peru places domesticated plants there at 6000 B.C. Previously the oldest known material from Peru came from the coast some two to three thousand years later. Gourds, squash, cotton, lima beans, and another pepper are the oldest plants known from the coast. Plants associated with human sites are also known from Mexico at 6000

B.C. or even earlier, but it is not clear that they were domesticated. However, by about 5000 B.C., gourds, squash, common beans, chili pepper, an amaranth, and avocado probably were domesticated. The gourds and common beans of Mexico are the same species as those found in Peru, but the squashes and chili pepper belong to different species.

Thus we have four general areas in the world where agriculture is quite old, and the archaeological record indicates that it is in the Near East that it is the earliest. As the people who see agriculture originating elsewhere point out, however, the archaeological record is never complete, and as Isaac puts it, the absence of archaeological support for the origin of agriculture in the tropics is not considered pertinent. It must be admitted that three of the areas, the Near East, Peru, and Mexico, where we point to early origins, are relatively dry, thus readily permitting plants to be preserved as artifacts.

As was pointed out above, Sauer concluded that vegetative agriculture developed before seed agriculture, and Isaac considers that tropical root and tuber cultivation is ancestral to seed cultivation, or developed independently of it. Again, however, from the archaeological record it is evident that seed agriculture is earlier in Mexico, Peru, and the Near East. All of the first plants from these areas are seed-propagated. The record from Southeast Asia is not yet clear, but it may prove that rice, another seed-propagated plant, is the earliest. Once again, however, perhaps the archaeological record is not "pertinent," for the dry seeds or fruits are much more likely to be preserved than are fleshy roots and tubers.

While I am willing to admit that vegetative planting may have been practiced as early as seed planting, it was the latter that led to the development of early high civilization in the Near East, Mexico, and Peru, and the same may yet prove true in Southeast Asia. Why seed cultivation led to the more rapid advances in human culture than did vegetative cultivation may stem from the more rapid evolutions that occurred in the seed crops. (One perhaps might also argue that a diet primarily of seeds is more nutritious than one composed primarily of roots and tubers.) Since seed crops reproduce sexually, and most of them are annuals, we might expect change to occur more rapidly in them than in the roots and tubers, which are usually propagated asexually. This would be likely particularly if intentional selection were practiced, and probable even if it was not. Sauer saw a distinction between seed and vegetative propagation in the New World, with seed cultivation predominant in Mexico and vegetative cultivation predominant in South America. Since that time the archaeologists have provided evidence that seed plants were domesticated very early in Peru. Although some of the major crops of South America—for example, manioc, sweet potato, and the Irish potato—are propagated vegetatively, we now know that there were also many indigenous domesticates grown from seed, among them beans, squash, quinua, and chocho. Although there can be little doubt that the Irish potato made major contributions to civilization in the very high Andes, I think the development would not have been as rapid without some of the seed crops, particularly quinua.

Whether seed or vegetative agriculture appeared first may never be settled to everyone's satisfaction. Similarly, it is hard to say whether there was a single origin of agriculture or multiple origins. If the question were put to a vote of anthropologists and botanists today, I think a clear majority would endorse the idea that there were multiple origins, but that, of course, does not mean that is necessarily the correct answer. The extreme diffusionists, including some geographers, hold to the view that there was but a single origin of agriculture followed by diffusion to other places. There is really no critical evidence to decide the issue, but I feel that multiple origins are more likely—or at least two origins, one in the Old World and one in the New. If, as the archaeological record indicates, agriculture first appeared in the Near East or

Egypt ten thousand years ago or earlier, there would probably be ample time for it to have spread to southeastern
Asia by the time agriculture appeared there. Likewise, we
might see agriculture originating either in Peru or Mexico
and then diffusing to the other areas in the New World.
To see agriculture coming from the Old World to the New,
however, calls for a long ocean voyage by people at a very
early date. Although we now know there were many voyages across the oceans to the Americas before Columbus,
it is unlikely that any of them go back to the time before
agriculture was known. Sometimes the fact that the plants
cultivated in early times in the New World and Old World
were completely different is used as evidence for separate
origins of agriculture in the two regions,[6] but the *idea*
for agriculture could have been carried from the Old
World to the New without any plants being involved.

   Was the concept of agriculture a difficult one for people?
If so, perhaps there were but one or two origins. On the
other hand, if it is a relatively simple concept, then perhaps
we can see many separate origins in various parts of the
world. There is no agreement, although today to many
people the idea of planting to secure more food does not
appear to be a particularly difficult one.

   Thus far I have been treating agriculture as a means of
producing food that could readily be recognized from the
archaeological record. When clearly domesticated plants
are found at a site, the conclusion is often reached that
the people had agriculture. This is not always justified,
for they could have had domesticated plants and still not
have been practicing agriculture. For that matter, people

----

   [6]An exception is the bottle gourd. Although almost certainly native
to Africa, it is found in the Americas by 7000 B.C. or earlier. However, since it has been shown that the fruits will float in ocean water
for a long period of time, it is not necessary to invoke people as the
agent of dispersal. The sweet potato was present in the Americas, where
it appears to be native, and several distant Pacific islands in pre-Columbian times. It is difficult not to invoke human aid to explain this
distribution.

could have had agriculture before there were any morphological changes in the plants; thus people could have been cultivating plants long before the archaeological record reveals any evidence of it. So far I have not defined agriculture. Generally it implies the field-scale cultivation of plants, as the derivation of the word indicates. It would seem obvious that field-scale cultivation must have been preceded by small-scale cultivation, that is, horticulture or gardening.

In some places, usually dating from much later than when agriculture is thought to have originated, we have archaeological evidence for it in the presence of seed storage bins, large-scale irrigation works, and, sometimes, the fields themselves, but from the remains of a few plants we cannot always tell whether horticulture or agriculture was being carried out. As should be quite obvious, it is not always possible to make a distinction between horticulture and agriculture. For example, how large does the field have to be before we speak of agriculture?

One thing that the archaeological record has revealed in Mexico, through the work of R. S. MacNeish and collaborators, is that when domesticated plants first appeared, they contributed only a small portion of the people's food, and it was only after a period of several thousand years that they became the dominant source of sustenance.

THE ORIGIN OF SEED PLANTING

One of the articles dealing with the origin of agriculture that I have found most enlightening is by Bennet Bronson. He makes very clear the distinction between the origin of planting and the origin of agriculture, and he maintains that the growing of useful plants was "neither a unique nor revolutionary event," that it probably happened in many places, and that its causes may have been comparatively trivial. It is some of these causes that I propose to examine, and obviously it will be speculation, for we can hardly expect the archaeological record to reveal them.

By far the simplest explanation is that people first made intentional plantings in order to secure more plants like the parent. Bronson considers it likely that the first plants grown were necessary or highly desirable and rare or scarce. Moreover, he thinks that it was not a major staple but more likely a fiber plant, a dye plant, or a stimulant. One does not have to be an archaeologist or a botanist to come up with a "hypothesis." For example, in his novel *The Source,* James Michener has the captive wife of Ur transplanting wheat to near her dwelling so her people would not have to search for it the next year. Apparently her father had done that before her. The origin of planting may be just that simple. I doubt, however, that wild plants were transplanted, for transplanting is often a difficult operation, and I would think it more likely that she scattered the grains where she wanted the plants. This "hypothesis," of course, is contrary to Bronson's for it involves a major staple, but it may be just as likely. All we have are guesses. It could well be that the origin of seed planting was not very complicated, but I have another hypothesis (or I would not have written this chapter)—namely that planting may not have arisen for economic reasons but had a religious motive.

Rites of first fruits were apparently widespread. Let us assume that at an early time some gatherers returned the seeds to the soil, not necessarily in the place where they were gathered but at some other site. Perhaps the seeds were actually buried, a more dramatic returning of them to mother earth, rather than simply scattered on the surface of the ground. The burial, of course, would have had to be shallow for the seeds to germinate. The area where this was done was remembered—or perhaps it was actually marked with stones as certain Arabs are known to have done in historical times. Thus the people would recognize the plants the next year as having come from the return of the first fruits. This would then be recognized as a sacred spot, and the first fruits from it would again be returned to the soil, but if there was more than enough to satisfy the gods, the remainder could have been used for food by the people. Thus the first planting and also the first sacred garden could have come into existence at the same time. The first sowing of seeds, therefore, would not have been an intentional planting but a return of the seeds to the gods so that more seeds would be forthcoming. A symbiosis between plants and human beings would have begun that would bring changes both to the plants and to the lives of the people.

Are we to explain all of seed planting as originating from first fruits? It depends on whether we look at it as originating only once or as having multiple origins. If it originated many times, then there may have been different causes in different places. If rites of first fruits were widespread among gatherers, however, perhaps events similar to the scenario developed above occurred more than once.

Although I have suggested a religious motivation for the original seed plantings, the further development of agriculture could well have been the result of secular acts, though one might also come up with religious motives to explain some of them. In any case, religion became attached to them soon after the act, as is apparent with irrigation in the Near East. Cultivating plants involves more than the planting of seeds: preparing the soil, weeding, and, in some places, watering. If somehow people realized that seeds should be buried in mother earth, this might have led to a crude preparation of the soil originally. One might suggest that if the plot from the planting of the first fruits was regarded as sacred, we might suppose that other plants were removed from it as they came up, and thus weeding would have originated. The knowledge that water produced a better growth was probably not difficult to acquire, but if the people thought that the sky god produced the "fertilizing element," rain, perhaps they duplicated his effort at times. Although I suppose that the first fruits were returned to the soil immediately following the harvest, we know that seeding was later postponed until spring or before the beginning of the rains. Perhaps some people found that they secured little germination when they made the planting immediately after the harvest and therefore followed it with a second planting at a later date.

We have to look to reasons why, once planting began, it should have been increased to give rise to agriculture. A number of different factors could have been responsible. Certainly population pressure might be one of them. If cities came into existence as trading centers before agriculture appeared, as some people have recently maintained, we might suppose that people who had already started planting increased the areas cultivated in order to have more food to barter with the traders. Perhaps there was no good reason at all for agriculture, but simply human propensity to carry out many activities beyond what is necessary, as seen in the overkill of animals, the increase in the size of temples, and the proliferation of university committees. In all probability there was no one factor that accounted for the emergence of agriculture, and it is probably futile to seek one. We can erect any number of hypotheses, some perhaps more reasonable than others, but proof may always be elusive.

The hypothesis that planting owes its origin to religion, however, is not as far-fetched as some would have it. Even though it may not have been the major factor, religion certainly played a role in the origin of agriculture. Agriculture and religion became intimately connected, and I think we can see agriculture playing a major role in the development of religion. The whole subject provides food for thought, and if the readers don't like any of the present hypotheses for the origin of agriculture, they can come up with their own. I only hope that they do not think I have led them down the garden path in vain.

Bibliographical Notes

CHAPTER 1

Conrad, N., D. L. Asch et al. 1983. *Prehistoric horticulture in Illinois: Accelerator radiocarbon dating of the evidence.* The University of Rochester Nuclear Structure Research Laboratory, Center for American Archaeology No. 54. Pp. 1–11.

Deltgen, Florian, and H. G. Kauer. 1973. The Claudius case. Harvard University Botanical Museum Leaflets 23: 213–44.

Girardot, Norman. 1983. *Myth and Meaning in Early Taoism.* Berkeley: University of California Press.

Gray, Asa. 1957. Naudin's researches into the specific characters and the varieties of the genus *Cucurbita. American Journal of the Sciences and Arts,* ser. 2, 24: 440–43.

Heiser, C. B. 1979. *The Gourd Book.* Norman: University of Oklahoma Press.

———. In press (1984). Some botanical considerations of the early domesticated plants north of Mexico, in *The Beginnings of Food Production in Prehistoric North America: The Archaeological Background and Botanical Evidence,* ed. Richard Ford. Ann Arbor: Museum of Anthropology, University of Michigan. This work is based on a conference held at the School of American Research, Santa Fe, New Mexico, in the spring of 1980.

———, and E. E. Schilling. In press (1984). Systematics of *Luffa:* A preliminary report, in *Proceedings of the Conferences on the Biology and Chemistry of the Cucurbitaceae,* ed. R. W. Robinson. Ithaca, N.Y.: Cornell University Press.

Norrman, Ralf, and Jon Haarberg. 1980. *Nature and Language: A Semiotic Study of Cucurbits in Literature.* Boston: Routledge and Kegan Paul.

King, Frances B. In press (1984). Early cultivated cucurbits in eastern North America, in *The Beginnings of Food Production in Prehistoric North America: the Archaeological Background and*

221

Botanical Evidence, ed. Richard Ford. Ann Arbor: Museum of Anthropology, University of Michigan.

Santino, Jack. 1983. Night of the wandering souls. *Natural History,* October pp. 43–50. An interesting account of Halloween customs with another version of the origin of Jack and the jack-o'-lantern.

Whitaker, T. W. No date. Cucurbitáceas americanas útiles al hombre. Provincia de Buenos Aires Comisión de Investigaciones Cientificas, pp. 1–42. Based on a talk given at La Plata, Argentina, in 1980. An excellent summary.

CHAPTER 2

Carter, George F. 1976. Chinese contacts with America: Fu-Sang again. *Anthropological Journal of Canada* 14: 10–24.

Forman, Sylvia Helen. 1977. The totora in Colta Lake: An object lesson in cultural change. *Ñawpa Pacha* 15: 111–16.

Heiser, C. B. 1978. The totora *(Scirpus californicus)* in Ecuador and Peru. *Economic Botany* 32: 222–36.

CHAPTER 3

Heiser, C. B. 1972. The relationships of the naranjilla, *Solanum quitoense.* Biotropica 4: 77–84.

———. In press (1984). A domesticated variety and relationships of *Solanum lasiocarpum,* in *The Biology and Systematics of the Solanaceae,* ed. William D'Arcy. New York: Columbia University Press.

Whalen, M. D., D. E. Costich, and C. B. Heiser. 1981. Taxonomy of *Solanum* section Lasiocarpa. *Gentes Herbarum* 12: 41–129.

CHAPTER 4

Asch, David L., and Nancy B. Asch. 1977. Chenopod as cultigen: A re-evaluation of some prehistoric collections. *Mid-Continental Journal of Archaeology* 2: 3–45.

Fernald, M. L., and A. C. Kinsey. 1958. *Edible Wild Plants of Eastern North America.* Revised by R. C. Rollins. New York: Harper and Row.

Gade, Daniel W. 1975. *Plants, Man and the Land in the Vilcanota Valley of Peru.* The Hague: W. Junk.

Simmonds, N. W. 1976. Quinoa and relatives, pp. 29–30, *Evolution of Crop Plants*, ed. N. W. Simmonds. London: Longman.
Wilson, Hugh. 1981. Domesticated *Chenopodium* of the Ozark Bluff Dwellers. *Economic Botany* 35: 233–39.
———, and C. B. Heiser. 1979. The origin and evolutionary relationships of 'Huauzontle' (*Chenopodium nuttalliae* Safford), domesticated chenopod of Mexico. *American Journal of Botany* 66: 198–206.

CHAPTER 5

Anon. 1983. Amaranth: The revival of an ancient crop. *The Cornucopia Project Newsletter* 3 (1):4. Published by the Regeneration Agriculture Association, Robert Rodale, President, 33 East Minor St., Emmaus, Pennsylvania 18049.
Coons, M. P. 1975. The genus *Amaranthus* in Ecuador. Ph.D. thesis, Indiana University, Bloomington.
———. 1982. Relationships of *Amaranthus caudatus*. *Economic Botany* 36: 129–46.
Guevara, Dario. 1960. Expresión ritual de comidas y bebidas ecuatorianas. *Humanitas* 11 (1): 1–57.
Haro, S. L. Alvear. 1976. Costumbres funerarias del Reino de Quito. *Boletín Instituto Panamerico de Geografia e Historia*, Nos. 13–14, pp. 1–24.
Hauptli, H., and S. Jain. In press (1984). Genetic structure of landrace populations of the New World grain amaranths. *Euphytica*.
Ruskin, F. R., ed. 1984. Amaranth: modern prospects for an ancient crop. Washington, D.C.: National Academy Press. I received a copy of this booklet sometime after I wrote this chapter. Not only is it beautifully illustrated, but also it contains considerable information about the plant and its history, as well as its potential.

CHAPTER 6

Heiser, C. B. 1963. A trip to Tulcán. *Bulletin of the Missouri Botanical Garden* 51 (9):3. See also the notes by E. A. (Edgar Anderson) in the same issue for his comments on páramos and topiary.
Thacker, Christopher. 1979. *The History of Gardens*. Berkeley: University of California Press.

CHAPTER 7

This chapter is based largely on an article written for the *Bulletin of the Missouri Botanical Garden* 52 (10): 8–13, at the request of Dr. Edgar Anderson. Additional references follow here.

Gade, D. W. 1969. Vanishing crops of traditional agriculture: The case of tarvi *(Lupinus mutabilis)*. *Proceedings of the Association of American Geographers* 1: 47–51.

Gladstones, J. S. 1980. Recent developments in the understanding, improvement, and use of lupines, pp. 603–11, in *Advances in Legume Science*, ed. R. J. Summerfield and A. H. Bunting. Kew: Royal Botanic Garden.

Gross, R., and E. Von Baer. 1982. The lupin—a new cultivated plant in the Andes, pp. 54–65, in *Plant Research and Development*. Tübingen: Institute for Scientific Cooperation.

CHAPTER 8

Anderson, Gregory J. 1979. Systematic and evolutionary consideration of species of *Solanum*, section *Basarthrum*. Pp. 549–62, in *The Biology and Taxonomy of the Solanaceae*, ed. J. G. Hawkes, R. N. Lester, and A. D. Skelding. London: Academic Press.

Heiser, C. B. 1964. Origin and variability of the pepino *(Solanum muricatum)*: A preliminary report. *Baileya* 12: 151–58.

Hudson, W. Donald. In press (1984). The relationship of wild and domesticated tomate, *Physalis philadelphica* Lamarck. In *The Biology and Systematics of the Solanaceae*, ed. William D'Arcy. New York: Columbia University Press.

CHAPTER 9

Heiser, C. B. 1976. *Capsicum*. Pp. 265–68, in *Evolution of Crop Plants*, ed. N. W. Simmonds. London: Longman.

———, and Barbara Pickersgill. 1969. Names for the cultivated *Capsicum* species (Solanaceae). *Taxon* 18: 277–83.

Jensen, R. J., M. J. McLeod, W. H. Eshbaugh, and S. I. Guttman. 1979. Numerical taxonomic analyses of allozymic variation in *Capsicum* (Solanaceae). *Taxon* 28: 315–27.

MacLeod, M. J., S. I. Guttman, W. H. Eshbaugh, and R. E.

Rayle. 1983. An electrophoretic study of evolution in *Capsicum* (Solanaceae). *Evolution* 37: 562–74.

Pickersgill, Barbara, C. B. Heiser, and J. McNeill. 1979. Numerical taxonomic studies of variation in some species of *Capsicum*. Pp. 679–700 in *The Biology and Taxonomy of the Solanaceae*, ed. J. W. Hawkes, R. N. Lester, and A. D. Skelding. London: Academic Press.

CHAPTER 10

Bailey, L. H. 1976. *Hortus Third*. New York: Macmillan.

Graf, A. B. 1982. *Exotica*, ser. 4. 2 vols. East Rutherford, N.J.: Roehrs.

Heiser, C. B. In press (1984). Topiary in Tulcán. *American Horticulturist*.

Trelease, W., and T. G. Yuncker. 1950. *The Piperaceae of Northern South America*. Urbana: University of Illinois Press.

The *Kew Index*, or more properly *Index Kewensis Plantarum Phanerogamarum*, which gives the names and places of publication of all species of seed plants, was originally published in two volumes by the Royal Botanic Gardens at Kew, England, in 1893–95. The compilation of this work was made possible by a gift from Charles Darwin. Since that time supplements have appeared at regular intervals. The work is indispensable to seed-plant taxonomists.

The number of species of *Peperomia* in cultivation has increased greatly in recent years. In his *Manual of Cultivated Plants* of 1949, Bailey lists only five species, whereas about fifty are given in *Hortus* III and about seventy-five in *Exotica*.

Many groups of plants — orchids, bromeliads, and African violets, for example — have societies devoted to them. I did not know of such for *Peperomia*, but my colleague, Charles Hagen, found the notice of one, and in view of my interest in them I have become a member. The Peperomia Society was founded in 1977, and the current membership chairman is Joni Vines, 311 East 11th Street, Hutchinson, Kansas 67501. Dues are five dollars a year. Their publication, the *Peperomia Society Gazette*, is issued quarterly. "*Peperomia serpens* 'Tena,' a new cultivar" appears in vol. 5 (1984), no. 4, pp. 10–11, 13.

CHAPTER 11

Asch, Nancy, and David Asch. 1978. The economic potential of *Iva annua* and its prehistoric importance in the lower Illinois Valley. Pp. 301–41, in *The Nature and Status of Ethnobotany*, ed. R. I. Ford et al. Ann Arbor, Mich.: University of Michigan Museum of Anthropology.

Jackson, Raymond C. 1960. A revision of the genus *Iva. University of Kansas Science Bulletin* 41: 793–876.

Yarnell, Richard. 1978. Domestication of sunflower and sumpweed in eastern North America. Pp. 289–99, in *The Nature and Status of Ethnobotany*, ed. R. I. Ford et al. Ann Arbor, Mich.: University of Michigan Museum of Anthropology.

CHAPTER 12

Gordon, Donald. 1966. A revision of the Genus *Gleditsia.* Ph.D. thesis, Indiana University, Bloomington.

Isley, Duane. 1975. *Leguminosae of the United States.* Vol. 2, *Subfamily Caesalpinioideae.* Memoirs of the New York Botanical Garden 25, no. 2.

McCoy, Scott. 1958. A new species of honey locust. *Proceedings of the Indiana Academy of Science* 68: 320–21.

Sargent, C. S. 1890. *Silva of North America,* vol. 3. Reprint, New York: Peter Smith, 1947.

CHAPTER 13

At one time I had thought of doing a whole book on the subject of religion and agriculture, but to handle the subject properly one would have to be a specialist in religion and mythology, which I certainly am not, as well as in botany and archaeology. So perhaps one day someone better qualified will undertake the task. My bibliography on this section runs to nearly two hundred entries, and I have chosen to cite only some of the more influential works. It will be obvious to the scholar that some of my references are secondary ones; many of the primary sources are in languages foreign to me, particularly German, of which in spite of my name I can claim no great knowledge.

The Rape of Persephone

The account of Demeter is condensed from the translation of Paul Friedrich, *The Meaning of Aphrodite* (Chicago: University of Chicago Press, 1978); see also for his commentaries. Other accounts, differing somewhat in details and interpretations, may be found in Robert Graves, *The Greek Myths* (New York: Braziller, 1957); Michael Grant, *Myths of the Greeks and Romans* (London: Werdenfeld and Nelson, 1962); P. O. Morford Mark and Robert J. Lenardon, *Classical Mythology* (New York: David McKay, 1971); and Meyer Reinhold, *Past and Present: The Continuity of Classical Myths* (Toronto: Hakkert, 1972).

The reference to the "thrice-plowed field" has been interpreted to mean that the Greeks actually plowed their fields three times before planting. This seemed unlikely to me, and I find that E. A. Armstrong ("The triple furrowed field," *Classical Review* 57 (1943: 3–5) maintains that there was a ritual plowing in which three furrows were turned at the beginning.

For the Eleusinian Mysteries, see George E. Mylonas, *Eleusis and the Eleusinian Mysteries* (Princeton, N.J.: Princeton University Press, 1961). The recent works referred to are C. Kerenyi, *Eleusis: Archetypal Image of Mother and Daughter* (New York: Schocken Books, 1977) and R. Gordon Wasson, C. A. P. Ruck, and Albert Hoffman, *The Road to Eleusis: Unveiling the Secrets of the Mysteries* (New York: Harcourt Brace Jovanovich, 1978). In the latter a mind-altering substance, ergot (growing on the grass *Lolium*), is identified as an ingredient in *kykeon*.

Frazer

Allen, Grant. 1894. The origin of cultivation. *The Fortnightly Review* 61: 578–92.

Bachofen, J. J. 1967. *Myth, Religion and Mother Right.* Bollingen Series 84. Trans. Ralph Manheim. Princeton, N.J.: Princeton University Press.

Downie, R. Angus. 1940. *James George Frazer: The Portrait of a Scholar.* London: Watts and Co.

Frazer, J. G. 1910 *Totemism and Exogamy.* New York: Macmillan.

Gaster, T. H., ed. 1959. *The New Golden Bough.* New York: Criterion Books.

For criticisms of Frazer, see Edmund Leach, et al., Frazer and Malanowski: A CA* discussion, *Current Anthropology* 7 (1966):

560–75. For an eloquent defense, see Robert Ackerman, J. G. Frazer revisited, *The American Scholar* 47 (1978): 232–35. Professor Ackerman has a biography of Frazer in preparation.

The Great Mother

Evans-Pritchard, E. E. 1965. *Theories of Primitive Religion*. London: Oxford.

Fairservis, Walter A., Jr. 1975. *The Threshold of Civilization: An Experiment in Prehistory*. New York: Scribner's.

James, E. O. 1959. *The Cult of the Mother-Goddess*. New York: Praeger.

Maringer, J. 1960. *The Gods of Prehistoric Man*. New York: Knopf.

Marshack, Alexander. 1972. *The Roots of Civilization*. New York: McGraw Hill.

Neuman, Erich. 1955. *The Great Mother: An Analysis of the Archetype*. Trans. Ralph Manheim. Princeton, N.J.: Princeton University Press.

Ochshorn, Judith. 1981. *The Female Experience and the Nature of the Divine*. Bloomington: Indiana University Press.

Sandars, N. K. 1979. The religious development of some early societies. Pp. 103–27, in *The Origins of Civilization*, ed. P. R. S. Moorey. Oxford: Clarendon Press.

Talbot, Percy A. 1927. *Some Nigerian Fertility Cults*. London: Oxford University Press.

Myths

Alexander, H. B. 1916. *North American Mythology*. Boston: Marshal Jones.

Eliade, Mircea. 1978. *A History of Religious Ideas*. Vol. I, *From the Stone Age to the Eleusinian Mysteries*. Chicago: University of Chicago Press.

Hatt, Gudmund. 1951. The corn mother in America and in Indonesia. *Anthropos* 46: 853–914.

Hooke, Samuel Henry. 1963. *Middle Eastern Mythology*. London: Penguin.

Isaac, Erich. 1970. *Geography of Domestication*. Englewood Cliffs, N.J.: Prentice-Hall. See also for his discussion of Jensen.

Jensen, Adolf E. 1963. *Myth and Cult Among Primitive People*. Trans. M. T. Choldin and W. Weissleter. Chicago: University of Chicago Press.

Kirk, G. S. 1970. *Myth, Its Meaning and Functions in Ancient and Other Cultures.* London: Cambridge University Press.

Mabuchi, Toichi. 1964. Tales concerning the origin of grains in the insular area of eastern and southeastern Asia. *Asian Folklore Studies* 23: 1–92.

Mark, P. O. Morford, and Robert J. Lenardon. 1971. *Classical Mythology.* New York: David McKay.

Reinhold, Meyer. 1972. *Past and Present: The Continuity of Classical Myths.* Toronto: Hakkert.

Sex and Sowing

Bachofen, J. J. 1967. *Myth, Religion and Mother Right.* Bollingen Series 84. Trans. Ralph Manheim. Princeton, N.J.: Princeton University Press.

Briffault, Robert. 1927. *The Mothers,* vol. 3. New York: Macmillan.

Hatt, Gudmund. 1951. The corn mother in America and in Indonesia. *Anthropos* 46: 853–914.

Kramer, Samuel N. 1969. *The Sacred Marriage Rite: Aspects of Faith, Myth and Ritual in Ancient Sumer.* Bloomington: Indiana University Press.

Talbot, Percy A. 1927. *Some Nigerian Fertility Cults.* London: Oxford University Press.

Westermarck, Edward A. 1922. *The History of Human Marriage,* Vol. 1. 5th ed. London: Macmillan.

Although I recall seeing a statement somewhere that references to sex in the fields are numerous, in spite of a rather diligent search, including a visit to the library of the Kinsey Institute for Research in Sex, Gender, and Reproduction, I have found very few documented examples other than those given in *The New Golden Bough.* The only place where I have found a larger number of cases brought together is in the most appropriately entitled book by Briffault.

Sacrifice

Eliade, M. 1954. *The Myth of the Eternal Return.* W. R. Traok. New York: Pantheon Books.

Hubert, Henri, and Mariel Mauss. 1964. *Sacrifice: Its Nature and*

Function. Trans. W. D. Halls. Chicago: University of Chicago Press.

Isaac, Erich. 1970. *Geography of Domestication.* Englewood Cliffs, N.J.: Prentice-Hall.

James, E. O. 1962. *Sacrifice and Sacrament.* New York: Barnes and Noble.

Jensen, Adolf E. 1963. *Myth and Cult Among Primitive People.* Trans. M. T. Choldin and W. Weissleter. Chicago: University of Chicago Press.

Neuman, Erich. 1955. *The Great Mother: An Analysis of the Archetype.* Trans. Ralph Manheim. Princeton, N.J.: Princeton University Press.

Sanday, Peggy R. 1981. *Female Power and Male Dominance: On the Origins of Sex and Inequality.* Cambridge: Cambridge University Press.

First Fruits and Sacred Gardens

Eliade, M. 1854. *The Myth of the Eternal Return.* Trans. W. R. Traok. New York: Pantheon Books.

Freeman, J. D. 1955. *Iban Agriculture.* London: H. M. Stationery Office.

Hatt, Gudmund. 1951. The corn mother in America and in Indonesia. *Anthropos* 46: 853–914.

Hubert, Henri, and Mariel Mauss. 1964. *Sacrifice: Its Nature and Function.* Trans. W. D. Halls. Chicago: University of Chicago Press.

James, E. O. 1962. *Sacrifice and Sacrament.* New York: Barnes and Noble.

Jenks, Albert E. 1900. The wild rice gatherers of the Upper Lakes. *Nineteenth Annual Report of the Bureau of American Ethnology,* part 2, pp. 1013–37.

Mabuchi, Toichi. 1964. Tales concerning the origin of grains in the insular area of eastern and southeastern Asia. *Asian Folklore Studies* 23: 1–92.

Sopher, David E. 1967. *Geography of Religions.* Englewood Cliffs, N.J.: Prentice-Hall.

Von Sydow, C. W. 1934. The Mannhardtian theories about the last sheaf and the fertility demons from a modern critical point of view. *Folklore* 45: 291–309.

The Origin of Agriculture

Allen, T. F. H., and H. H. Iltis. 1980. Overconnected collapse to higher levels: Urban and agricultural origins, a case study. Pp. 96–103, in *Systems Science and Science*. Proceedings of the Twenty-Fourth Annual North American Meeting of the Society for General Systems Research, with the American Association for the Advancement of Science. Ed. B. H. Banathy. Louisville, Ky.: Systems Science Institute.

Anderson, Edgar. 1952. *Plants, Man and Life*. Boston: Little, Brown.

Bronson, Bennet. 1975. The earliest farming: Demography as cause and consequence. Pp. 57–76, in *Population, Ecology and Social Evolution*, ed. Steven Polgar. The Hague: Mouton.

Bray, Warwick. 1976. From predation to production: The nature of the agricultural evolution in Mexico and Peru. Pp. 73–95, in *Problems in Economic and Social Archaeology*, ed. G. de G. Sieveking, I. H. Longworth, and K. E. Wilson. London: Duckworth.

Carter, George. 1977. A hypothesis suggesting a single origin of agriculture. Pp. 89–133, in *Origins of Agriculture*, ed. C. A. Reed. The Hague: Mouton.

Cohen, M. N. 1977. *The Food Crisis in Prehistory: Overpopulation and the Origins of Agriculture*. New Haven, Conn.: Yale University Press.

Evans-Pritchard, E. E. 1965. *Theories of Primitive Religion*. London: Oxford University Press.

Harlan, Jack R. 1975. *Crops and Man*. Madison, Wis.: American Society of Agronomy.

Heiser, C. B. 1981. *Seed to Civilization: The Story of Food*. 2d ed. San Francisco: W. H. Freeman.

Isaac, Erich. 1970. *Geography of Domestication*. Englewood Cliffs, N.J.: Prentice-Hall.

Rindos, David. 1980. Symbiosis, instability, and the origins and spread of agriculture: A new model. *Current Anthropology* 21: 751–72. See also *Current Anthropology* 22: 81–82, for additional criticism and discussion.

Runge, C. F., and D. W. Bromley. 1979. Property rights and the first economic revolution: The origins of agriculture reconsidered. Center for Resource Policy Studies, School of Natural Resources, Working paper no. 13. Madison, Wis.: College of

Agriculture and Life Sciences, University of Wisconsin.

Sauer, Carl O. 1952. *Agricultural Origins and Dispersals.* New York: American Geographical Society.

Zeven, A. C. 1973. Dr. Th. H. Engelbrecht's views on the origin of cultivated plants. *Euphytica* 22: 279–86.

The Origin of Seed Planting

Bronson, Bennet. 1975. The earliest farming: Demography as cause and consequence. Pp. 57–76, in *Population, Ecology and Social Evolution,* ed. Steven Polgar. The Hague: Mouton.

Heiser, C. B. 1981. *Seed to Civilization: The Story of Food.* 2d ed. San Francisco: W. H. Freeman.

My "first fruits" hypothesis for the origin of planting was first published in *Seed to Civilization* in 1973 (before I had read Frazer's *Totemism and Exogamy*). In the same year, *The First Farmers* by Jonathan N. Leonard (New York: Time-Life) appeared. In it the author mentions that a mythological origin of farming has long been out of fashion, but later he writes that when people "found patches of superior wheat springing up [from seeds that they had dropped] in their immediate neighborhood, they no doubt credited their gods with a benevolent miracle." Then after stating that deliberate planting could have originated in a number of ways, he goes on to say, "Perhaps grain was considered sacred and was scattered as an offering on the ground near some holy spot."

Index

233